TACKLING MARINE DEBRIS
IN THE 21ST CENTURY

Committee on the Effectiveness of International and National Measures to Prevent and Reduce Marine Debris and Its Impacts

Ocean Studies Board

Division on Earth and Life Studies

NATIONAL RESEARCH COUNCIL
OF THE NATIONAL ACADEMIES

THE NATIONAL ACADEMIES PRESS
Washington, D.C.
www.nap.edu

D1264432

THE NATIONAL ACADEMIES PRESS 500 Fifth Street, N.W. Washington, DC 20001

NOTICE: The project that is the subject of this report was approved by the Governing Board of the National Research Council, whose members are drawn from the councils of the National Academy of Sciences, the National Academy of Engineering, and the Institute of Medicine. The members of the committee responsible for the report were chosen for their special competences and with regard for appropriate balance.

This study was supported by Contract No. HSCG23-07-C-MMS158 between the National Academy of Sciences and the Department of Homeland Security. Any opinions, findings, conclusions, or recommendations expressed in this publication are those of the author(s) and do not necessarily reflect the views of the organizations or agencies that provided support for the project.

International Standard Book Number-13: 978-0-309-12697-7
International Standard Book Number-10: 0-309-12697-5

Cover: The front cover background image of debris on a beach was provided by Anthony F. Amos of the University of Texas Marine Science Institute. The image of the brown boobies on derelict fishing gear was provided by Dr. Dwayne Meadows of the National Oceanic and Atmospheric Administration. The images of the entangled Hawaiian monk seal and the entangled sea turtle were provided by the National Oceanic and Atmospheric Administration. The back cover and chapter-opening image of an abandoned fish trap was provided by Wolcott Henry 2005/Marine Photobank.

Additional copies of this report are available from the National Academies Press, 500 Fifth Street, N.W., Lockbox 285, Washington, DC 20055; (800) 624-6242 or (202) 334-3313 (in the Washington metropolitan area); Internet, http://www.nap.edu.

THE NATIONAL ACADEMIES
Advisers to the Nation on Science, Engineering, and Medicine

The **National Academy of Sciences** is a private, nonprofit, self-perpetuating society of distinguished scholars engaged in scientific and engineering research, dedicated to the furtherance of science and technology and to their use for the general welfare. Upon the authority of the charter granted to it by the Congress in 1863, the Academy has a mandate that requires it to advise the federal government on scientific and technical matters. Dr. Ralph J. Cicerone is president of the National Academy of Sciences.

The **National Academy of Engineering** was established in 1964, under the charter of the National Academy of Sciences, as a parallel organization of outstanding engineers. It is autonomous in its administration and in the selection of its members, sharing with the National Academy of Sciences the responsibility for advising the federal government. The National Academy of Engineering also sponsors engineering programs aimed at meeting national needs, encourages education and research, and recognizes the superior achievements of engineers. Dr. Charles M. Vest is president of the National Academy of Engineering.

The **Institute of Medicine** was established in 1970 by the National Academy of Sciences to secure the services of eminent members of appropriate professions in the examination of policy matters pertaining to the health of the public. The Institute acts under the responsibility given to the National Academy of Sciences by its congressional charter to be an adviser to the federal government and, upon its own initiative, to identify issues of medical care, research, and education. Dr. Harvey V. Fineberg is president of the Institute of Medicine.

The **National Research Council** was organized by the National Academy of Sciences in 1916 to associate the broad community of science and technology with the Academy's purposes of furthering knowledge and advising the federal government. Functioning in accordance with general policies determined by the Academy, the Council has become the principal operating agency of both the National Academy of Sciences and the National Academy of Engineering in providing services to the government, the public, and the scientific and engineering communities. The Council is administered jointly by both Academies and the Institute of Medicine. Dr. Ralph J. Cicerone and Dr. Charles M. Vest are chair and vice chair, respectively, of the National Research Council.

www.national-academies.org

COMMITTEE ON THE EFFECTIVENESS OF INTERNATIONAL AND NATIONAL MEASURES TO PREVENT AND REDUCE MARINE DEBRIS AND ITS IMPACTS

KEITH R. CRIDDLE (*Chair*), University of Alaska Fairbanks, Juneau
ANTHONY F. AMOS, University of Texas, Port Aransas
PAULA CARROLL, U.S. Coast Guard (retired), Honolulu, Hawaii
JAMES M. COE, National Oceanic and Atmospheric Administration (retired), Kirkland, Washington
MARY J. DONOHUE, University of Hawaii Sea Grant College Program, Honolulu
JUDITH H. HARRIS, Department of Ports and Transportation, City of Portland, Maine
KIHO KIM, American University, Washington, DC
ANTHONY MACDONALD, Monmouth University, West Long Branch, New Jersey
KATHY METCALF, Chamber of Shipping of America, Washington, DC
ALISON RIESER, University of Hawaii, Honolulu
NINA M. YOUNG, Consortium for Ocean Leadership, Washington, DC

Staff

SUSAN PARK, Program Officer
JODI BOSTROM, Associate Program Officer

Acknowledgments

This report was greatly enhanced by the participants of the three workshops held as part of this study. The committee would first like to acknowledge the efforts of those who gave presentations at meetings: Holly Bamford (National Oceanic and Atmospheric Administration), Nir Barnea (National Oceanic and Atmospheric Administration), Michael Blair (U.S. Coast Guard), Ginny Broadhurst (Northwest Straits Commission), Steve Collins (Cruise Lines International Association, Inc.), David Condino (U.S. Coast Guard), Charles (Bud) Darr (U.S. Coast Guard), Libby Etrie (U.S. Department of State), David Gravallese (Environmental Protection Agency), Andrew Gude (U.S. Fish and Wildlife Service), Martín Hall (Inter-American Tropical Tuna Commission), David Itano (University of Hawaii at Manoa), Jenna Jambeck (University of New Hampshire), Lindy Johnson (National Oceanic and Atmospheric Administration), Ilse Kiessling (Charles Darwin University), Bob King (Marine Conservation Alliance Foundation), Eric Kingma (West Pacific Regional Fisheries Management Council), Holly Koehler (U.S. Department of State), David Major (U.S. Coast Guard), Rene Mansho (Schnitzer Steel Hawaii Corporation), Thomas Matthews (Florida Fish and Wildlife Conservation Commission), Scott Muller (U.S. Coast Guard), William Nuckols (Coastal America), J. Michael Prince (University-National Oceanographic Laboratory System), David Redford (Environmental Protection Agency), Michael Simpkins (Marine Mammal Commission), Christine Ribic (University of Wisconsin), Seba Sheavly (Sheavly Consultants and The Ocean Conservancy), Rodney Smith (Covanta Energy), Mary Sohlberg (U.S. Coast

Guard), Heather St. Pierre (U.S. Coast Guard), Dick Stephenson (retired captain of tuna seiner *Connie Jean*), Paul Stocklin (U.S. Coast Guard), Michael Stone (Fury Group), Lisa Swanson (Matson Navigation Company), Steven Vanderkooy (Gulf States Marine Fisheries Commission), Howard Wiig (State of Hawaii), and Mark Young (U.S. Coast Guard). These talks helped set the stage for fruitful discussions in the closed sessions that followed.

This report has been reviewed in draft form by individuals chosen for their diverse perspectives and technical expertise, in accordance with procedures approved by the National Research Council's Report Review Committee. The purpose of this independent review is to provide candid and critical comments that will assist the institution in making its published report as sound as possible and to ensure that the report meets institutional standards for objectivity, evidence, and responsiveness to the study charge. The review comments and draft manuscript remain confidential to protect the integrity of the deliberative process. We wish to thank the following individuals for their participation in their review of this report:

Anne D. Aylward, U.S. Department of Transportation, Cambridge, Massachusetts

David Benton, Marine Conservation Alliance, Juneau, Alaska

Lillian C. Borrone, Port Authority of New York and New Jersey (retired), Avon, New Jersey

Russell E. Brainard, National Oceanic and Atmospheric Administration, Honolulu, Hawaii

Ginny Broadhurst, Northwest Straits Commission, Mount Vernon, Washington

Joseph T. DeAlteris, University of Rhode Island, Kingston

David G. Dickman, Venable, LLP, Washington, DC

Martín A. Hall, Inter-American Tropical Tuna Commission, La Jolla, California

Ilse Kiessling, Charles Darwin University, Darwin, Northern Territory, Australia

Judith E. McDowell, Woods Hole Oceanographic Institution, Massachusetts

Andrew A. Rosenberg, University of New Hampshire, Durham

Seba B. Sheavly, Sheavly Consultants, Virginia Beach, Virginia

Although the reviewers listed above have provided many constructive comments and suggestions, they were not asked to endorse the conclusions or recommendations nor did they see the final draft of the report before its release. The review of this report was overseen by **Andrew R.**

Solow, Woods Hole Oceanographic Institution, Massachusetts, appointed by the Divison on Earth and Life Studies, and **John E. Dowling**, Harvard University, Cambridge, Massachusetts, appointed by the Report Review Committee, who were responsible for making certain that an independent examination of this report was carried out in accordance with institutional procedures and that all review comments were carefully considered. Responsibility for the final content of this report rests entirely with the authoring committee and the institution.

Contents

xi

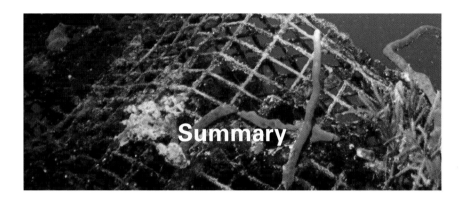

Summary

The debris of modern living frequently finds its way into our water-ways and down to the ocean. Some enters as intentional or accidental discharges from ships and platforms; the rest is transported to the sea by rivers, wind, sewers, and beachgoers. Given the diversity and abundance of sources, the persistent nature of most plastics, and the ability of tides and currents to carry debris long distances, marine debris is a global concern that is likely to increase in the 21st century.

The impacts of debris are varied. In 1988, it was estimated that New Jersey lost between $379 million and $3.6 billion in tourism and other revenue as a result of debris washing ashore. Impacts to marine organisms are often difficult to quantify but are well known. Ingested marine debris, particularly plastics, has been reported in necropsies of birds, turtles, marine mammals, fish, and squid. There is concern that plastics are able to adsorb, concentrate, and deliver toxic compounds to animals that ingest them. Derelict fishing gear (DFG) and other debris are known to entangle and injure or kill marine organisms. Studies on population-scale impacts of entanglement and ingestion are few and largely inconclusive. Nevertheless, these effects are troubling and may represent unacceptable threats to some species. For example, entanglement of Hawaiian monk seals, the most endangered seal in the United States, is arguably the most significant impediment to that species' recovery.

Marine debris regulation falls largely under the International Convention for the Prevention of Pollution from Ships, 1973, as modified by the Protocol of 1978 (MARPOL) Annex V, which entered into force in 1988.

This Convention places restrictions on the disposal of garbage, based on garbage type and distance from land, and completely prohibits the disposal of plastics at sea. Yet despite these and other prohibitions, 20 years later, there are still large quantities of debris, including plastics, fouling beaches and oceans.

STUDY ORIGIN

In 2006, Congress enacted the Marine Debris Research, Prevention, and Reduction Act. Its stated purposes are to identify, determine sources of, assess, reduce, and prevent marine debris and its impacts; revive interagency coordination efforts through an Interagency Marine Debris Coordinating Committee (IMDCC); and establish a federal clearinghouse for marine debris information. Within this Act, Congress requested that the National Research Council (NRC) undertake a study to assess the effectiveness of international and national measures to prevent and reduce marine debris and its impacts (see Box S.1 for the full statement of task).

Given its charge, the committee that wrote this report focused its efforts on the debris discharged at sea from a variety of maritime activities

---

BOX S.1
Statement of Task

An ad hoc committee will be formed to examine the effectiveness of international and national measures to prevent and reduce marine debris and its impact. The committee will prepare a report that includes

A. An evaluation of international and domestic implementation of MARPOL Annex V and the Act to Prevent Pollution from Ships (33 U.S.C. § 1901 et seq.) and identification of cost-effective, innovative approaches that could be taken to improve implementation and compliance.
B. A review and assessment of technologies, strategies, and management practices for further reducing the impact of marine debris, including derelict fishing gear. As part of this review, the committee will examine the International Maritime Organization's *Guidelines for the Implementation of Annex V of MARPOL* and recommend additional federal or international actions that could be taken to further reduce debris and its impacts.
C. An evaluation of the role of floating fish aggregation devices in the generation of marine debris and existing legal mechanisms to reduce impacts of such debris, focusing on impacts in the Western Pacific and Central Pacific regions.
D. An overview of the existing federal statutes on marine debris (including land-based sources) with a description of the responsibilities of the designated federal agencies.

including commercial shipping, fishing, recreational boating, and cruise ships. However, because it is unrealistic and impractical to differentiate between garbage discharged at sea and garbage that is discharged on land but winds up in the sea, this report addresses the ocean-based sources in the greater context of the marine debris problem. The committee recommends many specific actions that can be taken by decision makers and managers to spur a major paradigm shift toward a goal of "zero waste discharge" into the marine environment that the committee believes will be needed to effectively prevent and reduce marine debris. In addition, the report provides a specific review of DFG and abandoned fish aggregating devices (FADs).

Many of the recommendations in this report are not new. The ongoing problems with MARPOL Annex V and its implementation and recommendations for improvements were identified in the 1995 NRC report *Clean Ships, Clean Ports, Clean Oceans: Controlling Garbage and Plastic Wastes at Sea* and have been investigated in a number of other reports. However, a review by the U.S. General Accounting Office in 2000 found that many of these recommendations have not been fully implemented and some have not been implemented at all, indicating an ongoing problem with implementation of measures to prevent and reduce marine debris.

FINDINGS AND RECOMMENDATIONS

In its analysis, the committee identified four overarching areas in which additional emphasis is needed to adequately address the marine debris problem. Broadly, they center on (1) marine debris management, leadership, and coordination; (2) information and metrics with which to assess the effectiveness of current measures or efficiently direct future efforts; (3) port reception facilities for shoreside disposal; and (4) the distinct aspects of managing fishing gear as a source of marine debris. Salient supporting recommendations are presented under each overarching recommendation; additional recommendations and the basis for all of the findings and recommendations are included in Chapters 2, 3, and 4.

Management, Leadership, and Coordination

Despite measures to prevent and reduce marine debris, evidence shows that the problem continues and will likely worsen. This indicates that current measures for preventing and reducing marine debris are inadequate. Responsibilities and resources are scattered across organizations and management regimes, slowing progress on the problem. Improvements will require changes to the regulatory regime as well as nonregulatory incentives. At both the international and national levels, there needs to be better leadership, coordination, and integration of mandates and resources.

> **Overarching Recommendation:** The United States and the international maritime community should adopt a goal of zero discharge of waste into the marine environment. The United States should take the lead in the international arena in this effort and in coordinating regional management of marine debris with other coastal states. IMDCC should develop a strategic plan for domestic marine debris management. Performance measures should be developed by the United States and the international maritime community that allow for assessment of the effectiveness of current and future marine debris prevention and reduction measures (page 86).

Regulatory Structure

Under MARPOL Annex V, discharges are permitted unless specifically prohibited. This approach does not provide sufficient incentive to encourage innovation and adoption of source reduction and waste minimization measures to prevent garbage pollution in the marine environment.

> **Recommendation:** The U.S. delegation to the International Maritime Organization (IMO) should, through the ongoing review process, advocate that IMO amend MARPOL Annex V to include a general prohibition on discharge of garbage at sea with limited exceptions based on specific vessel operating scenarios and adequacy of shoreside reception facilities. In addition, the U.S. delegation should request that IMO review the *Guidelines for the Implementation of Annex V of MARPOL* and, where transferrable, amend MARPOL Annex V to include waste minimization and source reduction concepts from the *Guidelines* into mandatory requirements for vessels, such as within garbage management plan requirements. The United States and other parties to MARPOL Annex V should incorporate similar requirements into their domestic regulations for vessels engaged in both international and domestic trade (page 61).

Leadership and Coordination

Although Congress has charged federal agencies with addressing the marine debris problem and has called for interagency coordination, leadership and governance remain diffuse and ineffective. Current mitigation efforts are episodic and crisis driven. There is a need for a reliable, dedicated funding stream to support mitigation efforts and a national strategy and framework for identifying priorities for removal of marine debris.

> **Recommendation:** IMDCC or Congress should clearly designate a lead agency to expand cooperative marine debris programs, including but not limited to land-based marine debris, derelict fishing gear, shipborne waste, and abandoned vessels. IMDCC should develop a national strategy and national standards and priorities for dealing with all elements of marine debris. The strategic plan should include a clear identification of lead agencies, an implementation schedule, and performance benchmarks. In addition, IMDCC should identify funding mechanisms and reliable funding streams to support marine debris mitigation activities (pages 78 and 85).

Achieving Zero Discharge

There is a need to focus additional attention on potential waste before and after it reaches the ship. Zero discharge, source reduction, and waste minimization practices have been implemented in industrial settings ashore for a number of years. Some vessels have successfully adopted zero or minimal discharge practices based on these successful shoreside models.

> **Recommendation:** The U.S. Coast Guard (USCG), in coordination with the Environmental Protection Agency (EPA), should promulgate best management practices that reflect the maximum practicable extent to which ships can operate without the need to dispose of garbage at sea. Development of these best management practices should be based on successful zero discharge, source reduction, and waste minimization practices, coupled with an understanding of the technical and financial abilities of different vessel types to retain different forms of waste. IMDCC should support the adoption of voluntary zero waste discharge standards and implementation of these best management practices to achieve that goal. EPA should take the lead in coordinating with IMDCC to work with academia, industry, and nongovernmental organizations to develop industry standards and guidelines for source reduction, reuse, and recycling for solid wastes that are utilized and generated during normal ship operations (page 80).

Information and Metrics for Assessing Effectiveness

Although there is clear evidence that marine debris is a problem, there has not been a coordinated or targeted effort to thoroughly document and understand its sources, fates, and impacts. Mechanisms for objective evaluation are needed to judge the efficacy of management and mitigation measures; yet metrics for this evaluation are lacking. This confounds the ability to prioritize mitigation efforts and to assess the effectiveness of measures that have been implemented.

> **Overarching Recommendation:** IMDCC should, through planning and prioritization, target research to understand the sources, fates, and impacts of marine debris. It should support the establishment of scalable and statistically rigorous protocols that allow monitoring at a variety of temporal and spatial scales. These protocols should contain evaluative metrics that allow an assessment of progress in marine debris mitigation. The United States, through leadership in the international arena, should provide technical assistance and support for the establishment of additional monitoring and research programs worldwide (page 47).

Research

Diverse research has been conducted on marine debris; however, there is no overall needs assessment available to guide this research. As a result, research completed is rarely integrated at the regional, national, international, or even local level. Therefore, there is little opportunity for expanding the understanding of marine debris by fitting these individual activities into a congruous whole.

> **Recommendation:** An information needs assessment should be conducted at the national level by IMDCC with input from stakeholders. A detailed national marine debris research priorities plan should be developed from the results. This research plan should direct future federal funding of a suite of marine debris studies that, when taken together, will provide a comprehensive characterization of the marine debris problem. Additional studies are needed to assess the effectiveness of measures to prevent and reduce marine debris and to provide useful guidance to managers and decision makers for debris mitigation. IMDCC should sponsor and facilitate research in debris abundance and fluxes, and ecological and socioeconomic impacts of marine debris (pages 41 and 44).

Monitoring and Data Management

Well-designed and statistically rigorous longitudinal marine debris monitoring programs are needed at a variety of spatial and temporal scales. However, standardization of protocols and databases is necessary to ensure that the results of various surveys are comparable.

Recommendation: Long-term marine debris monitoring programs should be established by IMDCC (for the United States) and appropriate international organizations such as the United Nations Environment Programme (for global monitoring). These programs should allow for statistically valid analysis of marine debris quantities and trends as a metric of the effectiveness of measures to prevent and reduce marine debris. To the extent practical, these programs should adopt a suite of common design characteristics and protocols to facilitate cross comparisons and meta-analyses. The marine debris information clearinghouse should be given high priority. It should be housed and maintained by the National Oceanic and Atmospheric Administration (NOAA) but available to the public and researchers at large. Data generated by federally funded research should be submitted to this clearinghouse in a timely manner (pages 45 and 47).

Enforcement and Compliance Data

Forensic analysis of enforcement and compliance information is a necessary tool for evaluating the effectiveness of the implementation of MARPOL Annex V; however, there is no comprehensive system in place for collecting and analyzing information for this purpose at either the domestic or the international level.

Recommendation: USCG, in coordination with IMDCC, should develop a program to analyze the effectiveness of domestic regulations to reduce marine debris. Where feasible, it should utilize recordkeeping, enforcement, and other data that are already being collected and should investigate additional metrics that may be useful in measuring effectiveness. The U.S. delegation should recommend that IMO, in its ongoing review of MARPOL Annex V, incorporate this program into a global analysis of the effectiveness of MARPOL Annex V (page 83).

Port Reception Facilities

To prevent the discharge of waste at sea, ships must have the ability and incentives to properly dispose of waste onshore at port reception facilities. The lack of understanding of vessel waste streams and the inadequacy of port reception facilities to accept and properly manage vessel waste is a serious impediment to the prevention and reduction of marine debris, including DFG. Ships continue to face shoreside disposal challenges at some berths in countries that have formally communicated the availability of adequate reception facilities.

> **Overarching Recommendation:** To achieve the goal of zero discharge, ships need to be able to discharge their waste at ports and should have incentives (or at least they should not face disincentives) to do so. Domestically, USCG should establish minimum qualitative and quantitative standards for port adequacy, provide technical assistance for ports to achieve standards, encourage ports to provide incentives to vessel operators for discharging their waste ashore, and ensure that there are adequate reception facilities and alternative disposal options (see Appendix E) for waste fishing gear. Internationally, the U.S. delegation to IMO should exert its leadership in the ongoing MARPOL Annex V review process to ensure that similar amendments are incorporated into Annex V (page 86).

Regulatory Structure

While parties to MARPOL Annex V are required to ensure adequate port reception facilities, the standards for adequacy are unclear. Although the *Guidelines for the Implementation of Annex V of MARPOL* provides additional guidance, it does not establish minimum standards.

> **Recommendation:** The U.S. delegation to IMO should advocate that MARPOL Annex V be amended to include explicit qualitative and quantitative standards for adequate port reception facilities, and that IMO provide assistance to achieve these standards. Port managers and users should be included in the development of clearer standards. In addition, the U.S. delegation should encourage IMO to incorporate incentives for proper onshore waste disposal in these standards. In the United States, USCG should incorporate these minimum standards into their Certificate of Adequacy (COA) program and should encourage ports to provide incentives to vessel operators for discharging their waste ashore (page 63).

Integrated Solid Waste Management

Despite past recommendations and legislative mandates for collaboration, there continues to be a legal disconnect and jurisdictional discontinuity between solid waste management mandates afloat and ashore. There is no coordination between the Resource Conservation and Recovery Act of 1976 (RCRA), which regulates U.S. waste management and disposal, and the shipboard solid waste management plans or port and terminal waste management and COAs.

> **Recommendation:** Specific performance standards should be developed by USCG in collaboration with EPA for COAs; approval of port COAs should be conditioned on formal coordination between ports and solid waste management systems based on the RCRA waste management hierarchy and best management practices and guidance developed by EPA. Performance standards and COA and port discharge requirements should be based on an understanding of the capacity and capabilities of vessel types and waste streams, not just a hypothetical capability to handle wastes. The private sector and nongovernmental organizations should be included as partners in these efforts. EPA should work with state and local solid waste management programs and port and terminal operators to support a seamless connection and accountability for transfer of ship-generated garbage into the terrestrial waste management system (pages 78 and 79).

Managing Fishing Gear

While all maritime sectors contribute to ocean-based marine debris, there has been growing concern about the contribution of fishing vessels to this problem. Both DFG and FADs were specifically referenced in the Marine Debris Research, Prevention, and Reduction Act of 2006 as subjects for further review by this committee. DFG and abandoned FADs fall under MARPOL Annex V (and corresponding domestic laws) and fisheries management treaties and regulations. This overlap has complicated implementation of measures to prevent and reduce these sources of debris. Current regulations do not include accountability measures for gear loss, and fishermen and fisheries management organizations have few incentives and several disincentives to take responsibility for the impacts and for cleanup. Inadequate port facilities and high disposal costs are an impediment to the proper disposal of waste gear and DFG.

> **Overarching Recommendation:** MARPOL Annex V (and corresponding domestic law) and international and domestic fisheries treaties and regulations should be revised to clearly identify and prohibit preventable losses of fishing gear, including FADs. IMO, fisheries management councils and organizations, and other relevant entities should incorporate gear accountability measures and facilitate proper disposal of fishing gear, including FADs (page 140).

Regulatory Structure

MARPOL Annex V does not adequately manage discharges of fishing gear into the marine environment. The exemption for "the accidental loss of synthetic fishing nets, provided that all reasonable precautions have been taken to prevent such loss," does not provide sufficient guidance to regulators and the fishing industry. Moreover, because of minimum length and gross tonnage exemptions, MARPOL Annex V does not apply to a substantial number of fishing vessels; therefore, these vessels are exempt from many requirements that would facilitate enforcement of prohibitions against the at-sea disposal of synthetic fishing gear.

> **Recommendation:** The U.S. delegation should exercise its influence in the correspondence group and on IMO's Marine Environment Protection Committee to amend MARPOL Annex V to provide explicit definitions of "accidental losses" and "reasonable precautions" with respect to synthetic fishing nets; require placards, garbage management plans, and record books for all commercial, artisanal, and sport fishing charter vessels to the extent practicable; and require additional practices that minimize the probability of loss and maximize the probability of recovery of fishing gear from the ocean (page 102).

USCG and NOAA have rulemaking authority to prevent the generation of DFG under their respective legislative mandates, yet neither has exercised that authority.

> **Recommendation:** Congress should direct USCG and NOAA to undertake a joint rulemaking to develop rules that require commercial and recreational fishing vessels to properly dispose of all waste fishing gear and to take specific precautions to prevent accidental loss of fishing gear (page 106).

There has been confusion over the legal status of FADs in relation to marine debris. However, under MARPOL and Annex V definitions, FADs become DFG when the captain of the vessel that last deployed the FAD decides not to retrieve it. This constitutes an illegal disposal under MARPOL Annex V and the U.S. Act to Prevent Pollution from Ships (APPS) if the FAD includes synthetic ropes, webbing, or other plastics.

> **Recommendation:** NOAA should modify the federal regulations for U.S. tuna purse seine vessels to clarify the circumstances under which FADs become illegal discharges. Within international legal frameworks, the United States should encourage IMO and Regional Fisheries Management Organizations (RFMOs) to provide similarly

explicit definitions of "accidental losses" and "reasonable precautions" to clarify the circumstances under which FADs constitute illegal discharges of marine debris. RFMOs should devise regulations to exert greater control on the use, deployment, and retrieval of FADs to reduce the potential for FADs to become DFG. RFMOs should hold fishing fleets, nations, or the collection of all RFMO-licensed vessels responsible for retrieving all deployed FADs and should apply accountability measures such as loss of fishing privileges in RFMO waters. In turn, nations could potentially require retrieval of FADs by the vessel or fleet. In the United States, USCG should amend regulations implementing APPS to meet the intent of MARPOL Annex V and ensure that vessels fishing within U.S. waters and U.S. vessels fishing anywhere are held accountable to these standards (page 125).

Fisheries Management

The Magnuson-Stevens Fisheries Conservation and Management Act (MSFCMA)—the primary law governing U.S. fisheries management— does not highlight the need to reduce DFG or other fishery-related marine debris nor does it contain a national standard to address DFG or other marine debris. Although some Fishery Management Plans (FMPs) currently include measures that may have a collateral benefit of reducing DFG, current FMPs do not include measures that specifically address DFG.

Recommendation: Congress should add a national standard to MSFCMA that fishery conservation and management measures shall be designed to minimize the risk of gear loss. NOAA should establish a timetable for review of all existing FMPs for opportunities to reduce fishing-related marine debris, including reducing gear, minimizing gear loss, and minimizing impacts of lost gear, and to improve gear marking and recovery. Measures that reduce the loss or abandonment of fishing gear and encourage the retrieval of DFG should be considered in all future FMPs, National Environmental Policy Act documents, and the Endangered Species Act (ESA) Section 7 consultations and biological opinions. NOAA should encourage adoption of these measures by fisheries management organizations at the regional, state, and international levels. NOAA should also expand the duties of observers to include documentation of gear loss (page 115).

DFG has the potential to negatively impact endangered and protected species. For those fisheries that generate DFG that harms endangered and protected species, NOAA has the authority under ESA and the Marine

Mammal Protection Act (MMPA) to require fishing gear accountability measures.

> **Recommendation:** NOAA should determine which endangered and protected marine wildlife species or populations are at risk in part from DFG based on a review of all available information on fisheries interactions with these species, include information on injury and deaths due to DFG or other fishing-related marine debris in its marine mammal stock assessments and recovery plans and status reports for other threatened and endangered species, and use the provisions of ESA and MMPA to require adoption of gear accountability and other measures to minimize or remove DFG for fisheries that generate DFG that poses an entanglement threat to endangered and protected marine wildlife (page 116).

Currently, there is very little control or data on FADs in international fisheries. Replacement of plastic components and synthetic ropes and webbing used to construct FADs with readily degradable materials such as natural fibers would lessen the adverse impacts of FADs that become marine debris.

> **Recommendation:** The United States should take a leadership role by requiring that its own purse seine fleet submit a FAD management plan, encouraging RFMOs to adopt requirements for FAD management plans, and using port state jurisdiction in its territories to limit access to vessels flying the flag of countries that fail to require their vessels have a FAD management plan. RFMOs should control the number of FADs through chips, marking, tags, or other means to limit the number of FADs that can be carried and deployed by a vessel; acquire more information to characterize FAD usage in each of the agreement areas; adopt resolutions requiring parties to provide information on FAD use by vessel, including the number of sets on FADs, the number of FADs carried and deployed, and FAD retrieval, loss, and appropriation rates; and establish mechanisms to gather information on FADs including reports from parties, vessel logbooks, and observer programs. At a minimum, RFMOs need to collect and report annual data on the number of FADs deployed, the number returned to shore, the number lost, and an annual estimate of the number currently being fished. RFMOS should support the development of FAD designs that do not incorporate persistent synthetic or scrap materials but instead include materials that will self-destruct, readily biodegrade, mitigate entanglement, and provide an incentive for FADs to be maintained and regularly retrieved. RFMOs should

also prevent the use of synthetic and scrap material in FADs through regulation (pages 138 and 139).

International Cooperation

Because DFG persists and can be transported long distances, parties that generate DFG may not be the ones that bear the effects of it. Increased awareness and participation by responsible parties is necessary to effectively address the DFG problem.

Recommendation: All parties responsible for the generation of DFG should be involved in prevention and cleanup. Measures to prevent and reduce DFG will require international coordination and cooperation. NOAA, the U.S. Department of State, international fisheries management organizations, and other relevant organizations should engage in technology transfer and capacity building with nations from which DFG components originate to improve implementation of MARPOL Annex V in fisheries; encourage best practices to reduce gear loss, support recycling of used fishing gear, and promote retrieval of snagged or lost gear; and facilitate the participation of representatives from nations from which DFG components originate in DFG survey and removal efforts (page 92).

Gear Marking

Prevention of DFG begins at the source, but identifying the source may be difficult because ocean currents can transport DFG a long distance from the site of loss or discard and can involve substantial time lags. Effective gear marking is critical for identification of the sources of DFG and the fisheries that may have deployed this gear.

Recommendation: NOAA should convene a workshop to explore innovative and cost-effective approaches for identification or marking of trawls, seines, gillnets, longlines, and FADs to foster gear identification. Based on this information, NOAA should develop gear marking protocols that can be used in domestic and international fisheries to provide a structured basis for designing programs to reduce gear loss and abandonment and increase recovery of DFG (page 116).

Gear Loss, Recovery, and Disposal

Fishing is inherently hazardous and, of a necessity, entails some risk of gear loss despite all reasonable precautions. Because it is difficult for enforcement agencies to clearly differentiate between willful, preventable, and unpreventable gear losses, enforcement of a strict liability for gear losses would be problematic and could lead fishermen to underreport losses or obscure the location of gear losses.

> **Recommendation:** Fishery management organizations, if they adopt gear loss reporting and other accountability measures, should adopt a "no fault" policy regarding the documentation and recovery of lost fishing gear. Under this policy, local fishermen, state officials, and the public should work together to develop cost-effective DFG removal and disposal programs. These programs could be subsidized through user fees; a tax or deposit on trap tags, permits, or gear; public and private grants; or mitigation banking. Fishermen participating in removal efforts could receive financial credit, or at least be exempted from landfill tipping fees (page 118).

The high costs and difficulty in providing adequate reception facilities, particularly in remote areas, discourages proper disposal of used fishing gear and can also be a disincentive to DFG retrieval.

> **Recommendation:** The actual ability to receive used fishing gear and DFG should be incorporated into minimum standards in the assessment criteria for USCG COAs for port reception facilities. EPA, NOAA, and the U.S. Army Corps of Engineers, in cooperation with the fishing industry, ports, and fishery managers, should help fishing communities explore alternative strategies and technologies for management, disposal, and recycling of used and recovered DFG. IMDCC and the NOAA Marine Debris Program should consider expanding the marine debris cleanup grants program to help offset the disposal costs for recovered DFG. Consideration should be given to dropping the 50 percent match requirement for DFG recovery and disposal programs, particularly for small remote communities (page 119).

Some legal frameworks discourage or prevent the retrieval of DFG. In the United States, recovery of DFG may be inhibited by prohibitions against tampering with abandoned gear, the application of cabotage laws and burdensome certification requirements for vessels that transport DFG, and fishery regulations that prohibit vessels from carrying gear that is not a gear type permitted under their license endorsement.

> **Recommendation:** USCG should work with other federal agencies, state officials, fishermen, and the public to revise regulations that inhibit the removal of DFG (page 121).

It is immaterial whether the litter and other debris scattered along the shoreline or entangling marine animals was discarded from a vessel or discharged from a storm drain or whether or not the discharge was legally permitted. Although this report is focused on ocean-based debris sources, meaningful solutions will have to address the entire marine debris problem: the manufacture of materials that may become debris, the processes whereby debris is transported to the ocean, the organization of waste management and disposal systems, and the cleanup and remediation of regions that are impaired by marine debris. Progress will also require sustained funding and institutional support for the prevention and removal of marine debris. Even though the marine debris problem is international in scope, much could be done at the national, regional, state, and local levels. The United States, as a nation, can stop fouling its waters and the high seas and, in so doing, serve as a paragon for stewardship of the planet's defining ecosystem, the sea.

1
Introduction

The debris of modern living frequently finds its way into our waterways and down to the sea. Some debris enters the marine system as intentional or accidental discharges from ships and platforms; the rest is transported to the ocean by rivers, rain, wind, sewers, and beachgoers. Given the diversity and abundance of sources, the persistent nature of plastics[1] and other garbage, and the ability of tides and currents to carry debris long distances, marine debris is a global concern that is likely to increase in the 21st century. At the same time, marine debris is a problem that can, in part, be exacerbated or ameliorated by actions taken at local, state, regional, national, and international levels. This interplay between global and local dimensions of marine debris is an important attribute of the problem and its solutions.

For the purposes of this report, marine debris is defined as "any persistent, manufactured, or processed solid material that is directly or indirectly, intentionally or unintentionally, disposed of or abandoned into the marine environment."[2] This definition necessarily excludes natural

---

[1]The term "plastics" is used to encompass the wide range of synthetic polymeric materials that are characterized by their deformability and can thus be molded into a variety of three-dimensional shapes, including a variety of common materials such as polypropylene, polyethylene, polyvinyl chloride, polystyrene, nylon, and polycarbonate (National Research Council, 1994). Some plastics are degradable and not persistent.

[2]This definition was provided by the U.S. Coast Guard in the context of the committee's statement of task. However, it also closely follows the draft definition developed by the National Oceanic and Atmospheric Administration and the U.S. Coast Guard in response

flotsam, such as trees washed out to sea, and focuses on nondegradable synthetic materials that persist in the marine environment. Not all of these materials are inherently harmful, but evidence of damaging effects provides the impetus for this report, which focuses on measures to prevent and reduce the debris, particularly plastic debris, which has persistent negative impacts. It is also important to note that different types of marine debris have different effects. For example, a derelict net that is still actively ghost fishing raises concerns about entanglement of marine life, whereas a plastic water bottle discarded at sea may wash ashore and become a visual disamenity. As discussed in the following chapters, an improved understanding of the fates and impacts of various marine debris types will improve our efforts to prioritize mitigation.

Marine debris has many sources. Overall, most debris comes from land-based sources (e.g., household garbage, medical waste, plastic resin pellets used as inputs for plastics manufacturing), but a considerable amount of debris is discharged at sea[3] (e.g., U.S. Commission on Ocean Policy, 2004; Sheavly, 2007). Ocean-based sources of debris (e.g., fishing gear; galley waste; dunnage; cargo nets; wastes generated by offshore minerals and petroleum exploration, development, and extraction) may come from a diverse fleet of vessels and platforms. A 1995 National Research Council (NRC) report characterized 10 distinct U.S. maritime sectors: recreational boats; commercial fisheries; cargo ships; passenger day boats and ferries; small public vessels[4]; offshore platforms, rigs, and supply vessels; U.S. Navy combatant surface vessels; passenger cruise ships; research vessels; and miscellaneous vessels (National Research Council, 1995a). There are considerable differences between these sectors (e.g., number of vessels, average vessel size, average crew or passenger size, average time spent at sea), which can result in differences in garbage generation and waste management capabilities. In addition, these sectors are not static and there may be a great deal of variability within vessels of a single sector (see Box 1.1). Measures aimed at preventing and reducing marine debris will need to be tailored to the characteristics of each sector (National Research Council, 1995a).

Similarly, studies have shown significant regional differences in marine debris sources, abundance, impacts, and trends related to such factors as geographical location, oceanographic conditions, and proximity

---

to the Marine Debris Research, Prevention, and Reduction Act (33 U.S.C. § 1951 et seq.) ("Definition of Marine Debris for Purposes of the Marine Debris Research, Prevention, and Pollution Act," 73 Fed. Reg. 30322 [May 27, 2008]).

[3]The terms ocean-based, marine-based, shipborne, and maritime sources are used interchangeably to indicate marine debris that is discharged at sea.

[4]Includes small vessels belonging to the U.S. Coast Guard, the U.S. Navy, and other government entities.

BOX 1.1
Offshore Energy Development and Marine Debris

The offshore oil and gas sector illustrates the complex and evolving nature of ocean-based sources of marine debris. This sector encompasses a diversity of vessels, including fixed and floating offshore rigs and platforms, small service vessels, and seismic survey and exploration vessels. Marine debris from this sector ranges from seismic equipment that is lost or abandoned to debris that is accidentally or intentionally discharged during routine platform or support vessel operations. Some 10 percent of marine debris found on the Padre Island National Seashore in Texas has been attributed to offshore oil and gas operations (Miller and Jones, 2003). In 1994, "nearly all the offshore oil and gas exploration occur[red] in the [western] Gulf of Mexico" (National Research Council, 1995a) and, while this is still largely the case, recent interest in increased offshore drilling in U.S. waters (e.g., White House Office of the Press Secretary, 2008) could eventually lead to increased offshore exploration and production and associated marine debris, not only in the Gulf of Mexico but in coastal waters off Alaska, California, New England, and Florida.

to human populations (e.g., Ribic et al., 1992; Coe and Rogers, 1997; Donohue and Foley, 2007; Sheavly, 2007). As Box 1.1 indicates, the Gulf of Mexico coast is more vulnerable to marine debris from offshore oil and gas operations because of their prevalence in this region. In a recent national survey of coastal debris, it was found that land-based sources of debris dominate the region between Cape Cod, Massachusetts, and Beaufort, North Carolina, while ocean-based sources, particularly fishing gear, dominate the Hawaiian Islands (Sheavly, 2007). These differences are important when considering measures to prevent and reduce marine debris. There has been a movement toward regional approaches to address ocean and coastal problems at appropriate scales and to improve overall ocean governance (Joint Ocean Commission Initiative, 2007); given the regional differences in marine debris, this approach is relevant to marine debris management as well.

There are many points of intervention for addressing marine debris. Education can raise public awareness and change the behaviors leading to the discharge of waste that becomes marine debris. Prevention measures can also address waste production by minimizing the use of products that become marine debris or preventing them from being accidentally lost at sea. For maritime sources, there is a need for shipboard waste handling and storage options. Proper land-based waste disposal systems, including alternatives such as recycling, are also essential; however, along with adequate waste reception facilities, there is a need for incentives to use these facilities and disincentives for disposing of waste at sea. Finally, efforts can be directed at removing debris in the marine environment, and information on the scope of the marine debris problem will aid in prioritizing these removal programs.

This report focuses on the ocean-based sources of debris, but it recognizes that it is unrealistic and impractical to separate these sources in all situations and thus addresses the ocean-based sources in the greater context of the marine debris problem. In addition, this report provides a specific review of two marine debris types of increasing concern: derelict fishing gear (DFG) and abandoned fish aggregating devices (FADs).

MARINE DEBRIS TIMELINE

Humans once viewed the ocean and its resources as limitless and believed that disposal of waste from vessels and along rivers and coasts into the ocean would do little harm. However, awareness of marine debris as a significant waste management and ocean pollution problem has grown as more and more garbage, particularly persistent synthetic materials, has entered the marine environment. In the 1970s and 1980s, two major international conventions related to ocean garbage entered into force: the Convention on the Prevention of Marine Pollution by Dumping of Wastes and Other Matter, 1972 (commonly referred to as the London Convention) and the International Convention for the Prevention of Pollution from Ships, 1973, as modified by the Protocol of 1978 (MARPOL) Annex V.

During this same period, the marine debris problem was gaining attention in the United States as the public saw evidence of marine life entangled in debris as well as substantial amounts of garbage, including medical waste, washing up on beaches (e.g., Manheim, 1986; Adler, 1987; Toufexis, 1988). NRC first examined the problem of marine debris in the general context of ocean pollutants (National Research Council, 1975), but marine debris quickly came to be recognized as a problem in its own right. It was estimated that, in 1988, New Jersey lost between $379 million and $3.6 billion in tourism and other revenue as a result of debris washing ashore (Swanson et al., 1991; Ofiara and Brown, 1999). Concurrent losses in New York are estimated to have been between $950 million and $2 billion (Swanson et al., 1991). Public uproar over these wash-ups led, in 1989, to the development of the Floatables Action Plan for the New York Bight. From their inception through 2006, the Floatables Collection Programs have recovered over 353 million pounds of debris from the New York Bight, including more than 19.4 million pounds in 2006 alone (Environmental Protection Agency, 2007a). To address the growing threat to wildlife, the National Oceanic and Atmospheric Administration (NOAA) began the Marine Entanglement Research Program (MERP) in 1985. The Marine Plastic Pollution Research and Control Act of 1987 (33

U.S.C. § 1901 et seq.) established a marine debris coordinating committee, including senior officials from NOAA, the Environmental Protection Agency, the U.S. Coast Guard (USCG), and U.S. Navy. Against this backdrop, Congress was debating whether to phase out the continued ocean dumping of sewage sludge in waters off of New York and New Jersey. In 1995, NRC released *Clean Ships, Clean Ports, Clean Oceans: Controlling Garbage and Plastic Wastes at Sea* (National Research Council, 1995a), which included comprehensive recommendations to improve marine debris management. Similar recommendations are echoed in National Research Council (1996a) and Coe and Rogers (1997).

Despite the flurry and initial promise of these activities, many marine debris programs such as MERP received less support over the years or were entirely discontinued. In 2000, the U.S. General Accounting Office, in a report on reducing cruise ship pollution, reviewed the status of recommendations made in various studies aimed at strengthening U.S. enforcement efforts and discouraging illegal discharges. It found that most of the recommendations from the 1995 NRC report (National Research Council, 1995a) had been only partially implemented and some recommendations had not been implemented at all (U.S. General Accounting Office, 2000). The committee notes that most of these recommendations continue to be relevant and applicable today.

Awareness and concern about marine debris are once again on the rise. In 2004, the final report of the U.S. Commission on Ocean Policy, *An Ocean Blueprint for the 21st Century* set forth recommendations to improve efforts to assess the sources and consequences of marine debris; to reduce marine debris, including DFG; and to ensure the adequacy of reception facilities (U.S. Commission on Ocean Policy, 2004). Congress took action to address some of these recommendations and concerns in 2006 when it passed the Marine Debris Research, Prevention, and Reduction Act (MDRPRA) (33 U.S.C. § 1951 et seq.).

Several parallel activities have been spurred by this legislation and other concerns. MDRPRA reconstituted the Interagency Marine Debris Coordinating Committee, originally created by the Marine Plastic Pollution Research and Control Act, and charged it with coordinating "a comprehensive program of marine debris research and activities among federal agencies" (33 U.S.C. § 1951 et seq.). MDRPRA also legally established the NOAA Marine Debris Program, which had been relaunched by NOAA in 2005 to revive the work of MERP. NOAA's Marine Debris Program is actively involved in activities to study, prevent, and remediate marine debris (National Oceanic and Atmospheric Administration, 2008a). Internationally, the International Maritime Organization, the Asia-Pacific Economic Cooperation, the United Nations Environment Programme, the Food and Agriculture Organization of the United Nations, and the

International Oceanographic Commission, among others, are currently engaged in programs to improve the prevention and reduction of marine debris. Of particular interest is the ongoing work of a correspondence group of the International Maritime Organization Marine Environment Protection Committee "to develop the framework, method of work, and timetable for a comprehensive review of MARPOL Annex V *Regulations for the Prevention of Pollution by Garbage from Ships* and the associated *Revised Guidelines for the Implementation of MARPOL Annex V*" (International Maritime Organization, 2006a), which is discussed in further detail in Chapter 3. Given the fortuitous timing of these activities, the committee hopes that this report will provide useful input into the international review of MARPOL Annex V.

The recommendations of the U.S. Commission on Ocean Policy (2004) and the enactment of MDRPRA establish a good foundation for supporting more effective programs to reduce the amount and impact of marine debris from both ocean- and land-based sources. However, similar recommendations have been made before (e.g., Shomura and Godfrey, 1990; National Research Council, 1995a, 1996a; Coe and Rogers, 1997). Ongoing expansion of ocean-borne cargo transport and ever-increasing population density along U.S. and international coasts have the potential to overwhelm current marine debris management regimes. In moving forward, the committee has kept in mind the lessons learned from past attempts; this report outlines the committee's recommendations to further measures that will garner the ongoing support needed to address marine debris problems now and into the future.

STUDY APPROACH AND STATEMENT OF TASK

MDRPRA called for NRC to produce "a comprehensive report on the effectiveness of international and national measures to prevent and reduce marine debris and its impact" (33 U.S.C. § 1951 et seq.). USCG, as the study sponsor, worked with congressional staff and the Ocean Studies Board to refine the study charge (see Box 1.2 for the committee's full task statement).

The Committee on the Effectiveness of International and National Measures to Prevent and Reduce Marine Debris and Its Impacts was composed of experts with varying backgrounds and perspectives on the marine debris problem, from research to regulation, fisheries to shipping, and prevention and enforcement to impacts and mitigation. The committee met three times over the 15-month study period (December 17–18, 2007, in Washington, DC; February 20–22, 2008, in Irvine, California; and April 28–29, 2008, in Honolulu, Hawaii). At each of these meetings, there were public sessions during which the committee heard from federal

---

BOX 1.2
Statement of Task

An ad hoc committee will be formed to examine the effectiveness of international and national measures to prevent and reduce marine debris and its impact. The committee will prepare a report that includes

A. An evaluation of international and domestic implementation of MARPOL Annex V and the Act to Prevent Pollution from Ships (33 U.S.C. § 1901 et seq.) and identification of cost-effective, innovative approaches that could be taken to improve implementation and compliance.

B. A review and assessment of technologies, strategies, and management practices for further reducing the impact of marine debris, including derelict fishing gear. As part of this review, the committee will examine the International Maritime Organization's *Guidelines for the Implementation of Annex V of MARPOL* [International Maritime Organization, 2006b] and recommend additional federal or international actions that could be taken to further reduce debris and its impacts.

C. An evaluation of the role of floating fish aggregation devices in the generation of marine debris and existing legal mechanisms to reduce impacts of such debris, focusing on impacts in the Western Pacific and Central Pacific regions.

D. An overview of the existing federal statutes on marine debris (including land-based sources) with a description of the responsibilities of the designated federal agencies.

---

agency representatives, particularly members of the Interagency Marine Debris Coordinating Committee; marine debris and fisheries researchers; representatives from industry groups, fisheries management groups, and nongovernmental organizations; and many others (see acknowledgements for the full list of presenters). These presentations, as well as additional information submitted to or gathered by the committee throughout the study process, formed the basis for the findings and recommendations in this report. These findings and recommendations are supported by the best evidence available to the committee; however, in many cases, data for scientific assessment of the extent and impacts of marine debris are scarce. Therefore, many of the recommendations included in this report reflect the opinions and best judgment of the current committee as well as subjective judgments reflected in earlier reports. The committee's mandate was to evaluate measures to prevent and reduce ocean-based waste. Nevertheless, it is meaningless to artificially separate the land-based sources of marine debris from the discussion. Therefore, some of the discussion is on marine debris in general, regardless of source, and on options for changing the character and amount of materials entering into the waste stream and alternatives for waste disposal.

The report is intended as general guidance to U.S. policy makers and managers implementing measures to prevent and reduce marine debris. The U.S. federal government has the opportunity to demonstrate leadership in the global arena, while providing support and guidance for regional and local efforts within the country. Therefore, the actions taken by the U.S. federal government can improve marine debris mitigation measures at many levels, from international to local. At the same time, the committee believes that many of these recommendations are applicable at the state and local levels and also may be helpful to the governments of other nations struggling with the marine debris problem.

The committee chose to highlight broad principles, approaches, and technologies that are applicable across sectors and throughout the waste management process, though the report discusses in some detail FADs and other fishing gear. The report does not include a detailed review of sector-specific technologies or of the complex relationships between ports[5] and local waste handling systems and their fee structures. However, there is a special emphasis on the technologies and approaches available to the fishing industry and fishery-dependent communities with respect to the challenges of disposing of used fishing gear.

REPORT ORGANIZATION

Several previous and ongoing studies, particularly the 1995 NRC report (National Research Council, 1995a), have highlighted areas for improvement in national and international response to the marine debris problem. This report contributes to the ongoing dialogue by focusing on two overarching themes: a broad review of the effectiveness of MARPOL Annex V and its domestic implementation, and a specific look at the role of DFG and FADs as components of marine debris.

Chapter 2 includes a review of the available data on the quantity and impacts of marine debris in the environment, what these data reveal about efforts to prevent and reduce marine debris, and why additional and ongoing information is needed to support the development of a national strategy for addressing the source identification, prevention, mitigation, and remediation of marine debris, as well as to serve as a gauge of the effectiveness of the strategy.

Chapter 3 consists of a review and analysis of the existing regulatory and management framework for preventing and reducing marine debris

---

[5]The term "port" as used in this report is descriptive of both the harbor area where ships are docked and the agency (e.g., port authority or terminal operator) that administers the use of the public wharves and port properties (American Association of Port Authorities, 2006).

and what this information reveals about implementation of and compliance with MARPOL Annex V and complementary domestic statutes and regulations. In addition, Chapter 3 identifies gaps and flaws in the regulatory framework and its implementation and presents recommendations for addressing those shortcomings.

Chapter 4 includes a critical review of existing domestic and international laws as they relate to regulation of DFG and FADs and fishing practices that lead to the loss or abandonment of fishing gear. While DFG and abandoned or lost FADs are marine debris, there are many legal and practical aspects that make them unique from other types of debris and there is growing concern about their prevalence and impact. Moreover, DFG and FADs were specifically referenced in MDRPRA as subjects for further review by this committee. Therefore, the committee has devoted a separate chapter to exploring these types of debris.

The committee gathered a great deal of additional information that was relevant, but not central, to the study charge. Appendix C is a summary of selected data and literature on the quantities and impacts of marine debris, and Appendix D includes a list of parties to both MARPOL Annex V and members of one or more international fishing agreements. Appendix E, provided by Jenna Jambeck (Professor, Environmental Engineering, University of New Hampshire), describes in further detail the options available for recycling or disposing of used and abandoned fishing gear.

2

Understanding Marine Debris and Its Impacts

M arine debris presents a significant environmental challenge, far more diverse and less tractable than most other environmental issues. Marine debris, especially plastic debris, is now ubiquitous in the oceans and along coasts. It is found in the middle of the oceans (Matsumura and Nasu, 1997), on remote uninhabited tropical atolls (Donohue et al., 2001; McDermid and McMullen, 2004; Morishige et al., 2007), and on Arctic and sub-Antarctic islands (Gregory and Ryan, 1997). Despite heightened awareness of the problem and ongoing remediation efforts, studies suggest that, overall, marine debris in the environment has not been reduced (Miller and Jones, 2003; Barnes, 2005; Sheavly, 2007; Yamashita and Tanimura, 2007).

This chapter provides evidence for why marine debris is a serious and challenging problem. The body of work addressing marine debris is voluminous and an exhaustive literature review was not possible and not explicitly or implicitly part of the committee's statement of task. Instead, the committee summarizes selected peer-reviewed literature illustrating the prevalence and impacts of marine debris in the environment (see Appendix C) and assesses the effectiveness of measures to prevent and reduce marine debris based on this information. Knowledge gaps are identified and recommendations provided on key aspects of monitoring and research that can help improve assessments and prioritize marine debris mitigation efforts.

Much remains to be learned about marine debris sources, amounts, and impacts that will enhance efforts to reduce, prevent, and mitigate marine debris; however, existing information about marine debris and its impacts is sufficient to support immediate action to arrest this global environmental problem.

ABUNDANCE AND FLUX

For many people, the term "marine debris" evokes images of litter strewn on a beach, such as the one shown in Figure 2.1, but marine debris is much more than beach litter. Debris is found throughout the marine environment, from coastal waters to the deep ocean and from the sea surface down to the benthos (Figure 2.2). Monitoring is a primary mechanism for identifying sources; for understanding temporal and spatial trends in debris composition, prevalence, and distribution; and, therefore, for understanding the extent of the problem and the effectiveness of efforts to address it. The vastness of the world's oceans makes estimating the total amount of marine debris a significant challenge; nonetheless, the number and geographic coverage of studies carried out so far (see Appendix C, Tables I–III) highlight the worldwide pervasiveness of marine debris.

Coastal Environments

Coastal areas have served as the primary focal point for marine debris awareness and mitigation and remain hotspots of marine debris accumulation. Debris of both terrestrial and maritime origins converges and is concentrated at the land–sea interface. Because of the visibility of the problem, more is known about the occurrence and impact of marine debris along coastlines than in any other marine environment.

The prevalence of shoreline debris deposits is summarized in Appendix C, Table I. Marine debris items range from 4 to more than 48,000 items per kilometer (km) of shoreline, while the weight of the items ranges from 31 grams per km to more than 3.8 metric tons per km. Plastic materials dominate coastal marine debris in number, volume, and weight at all debris sizes examined to date, particularly on beaches and areas near population centers (e.g., Ribic et al., 1997; Sheavly, 2007). Because of the variation in methods used (e.g., data collected along transects from the waterline to the "edge" of the beach, along transects parallel to the shoreline, or along a strandline where debris is likely to be highest), straightforward comparisons among studies is problematic.

The majority of studies of coastal marine debris have noted increasing quantities of debris (e.g., Merrell, 1984; Ryan and Moloney, 1993; Walker et al., 1997; Willoughby et al., 1997; Velander and Mocogni, 1998), other studies found no change over time (e.g., Lucas, 1992; Williams and Tudor, 2001; Santos et al., 2005a; Sheavly, 2007), and a few studies have docu-

FIGURE 2.1 Image of a typical trash-covered beach (used with permission from the National Oceanic and Atmospheric Administration).

FIGURE 2.2 Spot prawn and rockfish swimming around a derelict commercial trap at 250 m depth off the central California coast (used with permission from Diana Watters, National Marine Fisheries Service, Southwest Fisheries Science Center, Fisheries Ecology Division).

mented decreases (e.g., Johnson, 1994; Edyvane et al., 2004). Teasing out the effects of regulatory changes in debris deposition can be difficult. For example, Johnson (1994) reported a decline in trawl webbing on a beach near Yakutat, Alaska, coincident with the implementation of the International Convention for the Prevention of Pollution from Ships, 1973, as modified by the Protocol of 1978 (MARPOL) Annex V. However, the pre– and post–Annex V periods examined correspond to the transition from foreign to joint venture to domestic fisheries in the Gulf of Alaska. Moreover, during this same period, Gulf of Alaska groundfish catches dropped by more than 50 percent; thus, the observed decline in trawl webbing could be due to MARPOL Annex V, a change in fishing intensity, or a combination of factors. Elucidating the contribution of mitigation actions on the abundance of marine debris can only be achieved when there is an opportunity to draw on a series of compatible longitudinal surveys.

In the United States, after ratification of MARPOL Annex V in 1988, efforts were made to document any changes in marine debris as indicated by accumulations on beaches. In 1989, these efforts led to the establishment of the National Marine Debris Monitoring Program (NMDMP) (Sheavly, 2007). Following the congressional ratification of MARPOL Annex V and the enactment of the Marine Plastic Pollution Research and Control Act (33 U.S.C. § 1901 et seq.), an interagency workgroup (including the Environmental Protection Agency, the National Oceanic and Atmospheric Administration [NOAA], the National Park Service, and the U.S. Coast Guard) designed NMDMP. The aim of NMDMP was to evaluate the magnitude of the marine debris problem, its geographical distribution, any seasonal or long-term trends, and debris sources, and to do so according to a statistical design and sampling protocol that would allow for comparisons through time and across regions nationwide (Escardó-Boomsma et al., 1995).

NMDMP used indicator items to assign debris to presumptive ocean- or land-based sources, as well as to quantify items of particular concern. The Center for Marine Conservation (now known as The Ocean Conservancy) conducted the monitoring from 1996 to 2006 (data used for the five-year national analysis was collected from 2001 to 2006) and released their findings in a 2007 report (Sheavly, 2007). The study shows that, for the nation as a whole, there has been no statistically significant change in the prevalence of marine debris. Of the nine regions surveyed, only the Hawaiian Islands showed a significant decrease, and the effects of El Niño may have been a contributing factor (see Sheavly [2007] for a detailed summary of results). That is, NMDMP results indicate that the accumulation of litter on the nation's beaches is not diminishing.

Pelagic Environments

Since Heyerdahl's (1970) report on the occurrence of marine debris in the open ocean, a growing number of studies have provided a greater but incomplete understanding of the problem. These studies reveal that estimates of marine debris in the near-surface zone are highly variable, encompassing five orders of magnitude from less than 1 item per square km to as many as 332,556 items (about 5 kg) per square km (see Appendix C, Table II). Reported variability in debris prevalence likely reflects differences in sampling techniques, total area sampled, geographic locations, sea states, and timing of the sampling exercises, as well as non-uniformities in the distribution of debris. For instance, studies using nets to sample pelagic debris have employed net mesh sizes ranging from 150 to 947 μm, whereas visual shipboard surveys have detection limits on the order of several centimeters. Selection of monitoring areas also has a clear impact on sampling results. The comparatively high density of marine debris documented by Moore et al. (2001a) reflects the selective sampling of an ocean surface convergence zone where debris is known to accumulate (Matsumura and Nasu, 1997; Pichel et al., 2007). Similarly, work by Yamashita and Tanimura (2007) revealed an extremely patchy distribution of plastic pellets in the Pacific Ocean near Japan. One sampling tow showed upwards of 174,355 items per square km, whereas 21 of 76 tows (28 percent) contained no plastic pellets whatsoever. Because these studies did not examine pelagic debris distributions before and after implementation of MARPOL Annex V, there is no information to determine if MARPOL Annex V has contributed to demonstrable changes in the prevalence of pelagic marine debris.

Benthic Environments

Although less readily observed than marine debris in coastal or pelagic environments, marine debris is also present on the sea floor (see Appendix C, Table III). However, given the difficulties of sampling benthic environments, there have only been a small number of studies of benthic debris and those noted herein have focused on areas that are shallower than several hundred meters in depth. Generally, these studies show that benthic debris density is positively correlated with proximity to human activity (i.e., greater debris density nearshore vs. offshore). For instance, submerged areas adjacent to beaches in Curaçao (Nagelkerken et al., 2001) and Indonesia (Uneputty and Evans, 1997) have the two highest reported densities of benthic marine debris. However, distance from large population centers does not provide a uniform guarantee of low debris levels; Donohue et al. (2001) documented the presence of marine debris in shallow waters of the uninhabited islands of the Northwestern Hawaiian Islands (NWHI) and concluded that fishing gear fouling the benthos of these islands originated from distant water fisheries.

Debris Fluxes

Buoyant materials introduced at sea are either entrained along convergence zones or move toward shore, albeit over long time scales, from years to decades (Kubota, 1994; Donohue, 2005; Pichel et al., 2007). Floating marine debris is known to form dynamic patches and aggregations, at least in the Pacific Ocean (Kubota, 1994; Ingraham and Ebbesmeyer, 2001; Kubota et al., 2005). Storms and spring tides have been observed to uncover and resuspend debris that has been incorporated into shoreline sediments (Johnson and Eiler, 1999). The rate at which debris in the open ocean is entrained toward the shoreline, and the frequency with which debris is moved from the shoreline into the open seas and redeposited on the shore, is poorly understood (see National Research Council [1995a] for discussion). Anecdotal evidence from debris cleanup activities suggests that debris may remain in the pelagic realm for extended periods. Understanding these lag processes is an important element in predicting the amount of time that may be required to detect changes in the quantity of debris introduced into the marine environment and thus the effectiveness of management measures intended to reduce debris discharges.

An area that has received little attention is the understanding of factors that affect and promote the vertical transport of debris. While the fates of dense materials (e.g., metal, glass) are clear, vertical movement of plastics appears to be more complicated. Understanding vertical transport of plastic will require an understanding of biophysical and chemical processes that contribute to its breakdown and affect its buoyancy (Hollstrom, 1975; Ye and Andrady, 1991). Many, but not all, plastics are neutral to positively buoyant and thus remain on or near the ocean surface. Nylons, aramids, and many carbon fiber compounds used as high-tensile cording are neutral to negatively buoyant and sink to the benthos. Moreover, through photodegradation, mechanical breakdown, or fouling with organic matter, even buoyant plastic debris sinks or is transported to the benthos. However, the long-term fate of marine debris on the benthos is unclear as erstwhile buoyant debris has been observed to become unfouled and return to the surface (Ye and Andrady, 1991). Given the difficulty of sampling the seabed, understanding both the dynamics of the vertical transport of plastics as well as the degradation of plastics at different depths would be useful in understanding the full extent of the marine debris problem in the oceans.

IMPACTS

Understanding the impacts of different types of marine debris is as important as understanding its temporal and spatial prevalence. Not all types of debris are equally harmful, and not all organisms or regions are equally vulnerable. To prudently use scarce or limited resources in mitigation efforts, it is important to fully understand the impacts of marine debris on the environment and on human uses. Horrific images of seabirds, turtles, and marine mammals, dead and dying as a result of ingesting or becoming entangled in debris, have often been the public image of the marine debris problem. This section illustrates the serious consequences of marine debris in the environment, beginning with the ecological implications and concluding with a discussion of some of the socioeconomic impacts. Although understanding of the full breadth of impacts is far from complete, it is nonetheless clear that enough is known to warrant action.

Ingestion

Ingested marine debris, particularly plastics, has been reported in necropsies of birds, turtles, marine mammals, fish, and squid (Laist, 1997). A review of the literature[1] indicates that ingested marine debris is quite common in samples of dead and captured seabirds and turtles (see Appendix C, Table IV). The large variation in the prevalence of ingested debris does not appear to correlate with any particular taxa, region, or time period. However, there may be regional trends. For example, Robards et al. (1995) reported that the number of plastic particles ingested by some seabird species has increased over time in the subarctic waters off of Alaska.

The known effects of ingestion of marine debris by birds include reducing the absorption of nutrients in the gut, reducing the amount of space for food in the gizzard and stomach, uptake of toxic substances that comprise the debris or have been adsorbed onto the debris, ulceration of tissues, and mechanical blockage of digestive processes (Azzarello and Van Vleet, 1987; Fry et al., 1987; Ryan and Jackson, 1987; Ryan, 1988; Spear et al., 1995; see also Table IV in Appendix C).

Prevalence of debris ingestion among seabirds is suggestive of a broad and significant ecological impact, at least in some regions such as the North Pacific Ocean. However, a direct link between ingestion and mortality has been limited to incidental examinations of a small number of birds (Pierce et al., 2004). Spear et al. (1995) found a statistically significant positive correlation between body weight and the presence of plastic particles and a statistically significant negative correlation between the number of plastic particles ingested and body weight. Other studies have failed to detect a statistically significant correlation between plastic ingestion and body condition (Furness, 1985; Sileo et al., 1990; Moser and Lee, 1992; Shaw and Day, 1994; Vlietstra and Parga, 2002).

In addition to the possible direct physical effects of marine debris, there is concern that plastics, particularly microplastics, are able to adsorb, concentrate, and deliver toxic compounds to organisms that ingest them or to benthic communities. Microplastics are the very small (approximately ≤5 mm) plastic debris items; sources include preproduction plastic resin pellets used in the manufacture of plastic items (Gregory, 1977, 1978; Shiber, 1979, 1982, 1987; Redford et al., 1997; Moore et al., 2001b), tiny bead "scrubbers" used in washing products (Zitko and Hanlon, 1991; Gregory, 1996), abrasive plastic beads used to clean ships (Reddy et al., 2006), and ever-smaller fragments resulting from the mechanical and photodegra-

---

[1]Only publications citing samples sizes greater than 20 individuals examined were included in this review. The literature includes many additional reports of ingestion by seabirds, turtles, marine mammals, and fish, but these are generally limited observations that are not suitable for estimating population-level frequencies.

dation (oxidation) of larger plastic debris (Andrady, 1990; George, 1995). Plastic resin pellets have been shown to adsorb hydrophobic organic contaminants including polychlorinated biphenyls (PCBs), dichlorodiphenyl-dichloroethylene (DDE), and nonylphenols (Mato et al., 2001; Endo et al., 2005; Rios et al., 2007). Mato et al. (2001) suggested that contaminated plastic particles may serve as a source of toxins to organisms that ingest them. Teuten et al. (2007) showed that sorption of contaminants to plastics greatly exceeded that to two natural sediments, and that as little as 1 μg of polyethylene contaminated with phenanthrene per gram of sediment significantly increased the accumulation of this contaminant by an invertebrate worm; they postulated that plastic may serve to transport hydrophobic contaminants to sediment-dwelling organisms at the base of the food chain and as such provide a mechanism for amplification of contaminants throughout the food web (e.g., Gregory, 1996).

The small size of microplastic marine debris allows it to be ingested by a wide range of organisms. Microplastic particles as small as 20 μm can be ingested by invertebrates, including lugworms, barnacles, and amphipods (Thompson et al., 2004), and by protochordates such as salps (Moore et al., 2001a). The ingestion of plastic particles by seabirds and marine mammals has also been widely reported (Fry et al., 1987; Moser and Lee, 1992; Laist, 1997; Robards et al., 1997). Moreover, the correlation between toxic load and amount of plastic ingestion in seabirds has been known for two decades (Ryan et al., 1988). Toxic effects would be expected to compound damage resulting from the suite of known impacts related to the ingestion of plastics by marine birds.

Although the effects of plastic ingestion may not currently rise to the level of significantly impacting population-scale dynamics, ingestion of plastic debris may impede the recovery of species listed under the Endangered Species Act (16 U.S.C. § 1531 et seq.). Furthermore, ingestion-related injuries and mortalities, even those that do not threaten populations, may evoke substantial public concern.

Entanglement

The effect on organisms of entanglement in marine debris ranges from restricting the movement of affected individuals to direct physical harm and mortality. Sessile animals are not immune from what could be considered a form of entanglement via the scouring, abrading, or breakage they experience, as in the case of live coral reefs, when marine debris snags or entangles them (as discussed later under "Other Ecological Impacts"). Although entanglement morbidity and mortality of individual animals is of concern, the potential effect of entanglement on animal populations is also of conservation and legal interest.

Three pieces of information are needed to understand the population-scale impacts of entanglement: entanglement rate, entanglement-related mortality rate, and the demographic structure of the species or population under study. Of the many studies reporting on debris entanglement, few provide entanglement rates, fewer provide mortality rates, and only a handful report mortality rates in the context of populations (see Appendix C, Table V).

Most quantitative studies of debris-related entanglements have focused on marine mammals, birds, and turtles (see Appendix C, Table V). The prevalence of entanglement (number of cases per population) was generally less than 1 percent; however, entangled animals may die unobserved at sea or otherwise fail to return to land after entanglement, confounding both entanglement rates and subsequent fate of entangled animals (Laist, 1997). Entanglements typically involve debris that encircles the neck or appendages, most commonly plastic packing straps, followed by rope and line, and net fragments (Laist, 1997; Henderson, 2001). Once entangled, mortality rates differ among species from more than 80 percent for Antarctic fur seals, 44 percent for Australian sea lions, and 57 percent for entangled New Zealand fur seals (Croxall et al., 1990; Page et al., 2004). Entanglement of the most endangered seal in the United States, the Hawaiian monk seal, is arguably the most significant documented impediment to the species' recovery (Boland and Donohue, 2003), with a mean annual population entanglement rate of 0.70 percent reported from 1982 to 1998 (Henderson, 2001).

In addition to entanglement rates and entanglement-induced mortality rates, whether entanglement poses a significant threat to a species or stock is dependent on the demographic structure of the species or population (e.g., population growth rates). For instance, Fowler (1987) suggested that even though the entanglement rate for northern fur seals in the Pribilof Islands, Alaska, was less than 0.5 percent, an estimated entanglement-related mortality rate among juveniles of 15 percent is thought to have contributed to the decline in this species, which is now listed as depleted under the Marine Mammal Protection Act of 1972 (16 U.S.C. § 1361 et seq.). An additional example is the critically endangered Hawaiian monk seal. With 1,250 individuals remaining (Carretta et al., 2007), the success of juvenile recruitment is key to the species' survival. Juvenile Hawaiian monk seals have been shown to become entangled more frequently than adults (Henderson, 2001), hampering the species' recovery. For other species reviewed here, entanglement has not been found to be an important factor in the current status of their populations (Croxall et al., 1990; Arnould and Croxall, 1995; Zavala-González and Mellink, 1997; Hanni and Pyle, 2000; Hofmeyr and Bester, 2002; Hofmeyr et al., 2006).

As with debris ingestion, even if entanglement-induced mortality does not rise to a level that significantly impacts population viability, entanglement may impede the recovery of species listed under the Endangered Species Act and may evoke substantial public concern. There is no clear temporal trend in entanglement rates among marine mammals as a group (see Appendix C, Table V). While some studies have documented declines in entanglement rates that may be attributed to implementation of MARPOL Annex V (e.g., Antarctic fur seal as reported by Arnould and Croxall [1995]), others have reported increased entanglement rates (Zavala-González and Mellink, 1997; Page et al., 2004). The one study specifically evaluating the impact of MARPOL Annex V found no post-implementation abatement of entanglement of the endangered Hawaiian monk seal (Henderson, 2001). Moreover, separating the effect of MARPOL Annex V from the background of other regulatory and institutional changes can be problematic. For example, reported changes in northern fur seal entanglement rates (e.g., Fowler and Baba, 1991) did not account for the effects of radical changes in the structure and organization of the Bering Sea fisheries that took place during the 1980s and 1990s that may have been the ultimate cause of the variations observed in northern fur seal entanglements.

Ghost Fishing

Ghost fishing is a widely acknowledged but poorly understood problem of derelict fishing gear (DFG). To fully comprehend the magnitude of the impact, a number of parameters must be determined, including the amount of lost gear, catch or mortality rates, the length of time the gear continues to actively fish, and the dynamics and demographics of the populations of fish and shellfish captured in the gear (Breen, 1987). Because these parameters vary by fishery and even by location for a given fishery, the biological and economic impacts of ghost fishing are difficult to quantify. Additional discussion of the characteristics of fishing gear types, causes of gear loss, and impacts is included in Chapter 4.

Ghost fishing is primarily a problem associated with static gear such as gillnets, hook-and-line, traps, cages, and pots rather than active gear such as seines and trawls. Studies show that gillnets, once lost, can continue to actively fish for some time (see Appendix C, Table VI). While ghost nets can cause substantial mortality, estimates suggest that ghost fishing mortality in gillnets and tangle nets is a small fraction of directed catches. For example, Sancho et al. (2003) and Brown and Macfayden (2007) estimated that ghost fishing losses do not exceed 5 percent of commercial landings in European gillnet and tangle net fisheries. However, in the case of small stocks, ghost fishing mortality may be a cause for conservation concern. For example, Kappenman and Parker (2007) reported that ghost gillnets may continue to be active for as long as 7 years and are estimated to account for annual losses of 545 white sturgeon (*Acipenser transmontanus*) in the Columbia River, a mortality rate that is approximately one-third the magnitude of the commercial harvest. As nets become fouled (i.e., become more visible) or collapse due to initial capture, ghost fishing capacity declines. In contrast, for traps, cages, and pots, which are used to target crustaceans and some species of finfish, ghost fishing can persist for as long as the gear remains intact. Thus, a key component of understanding the impact of derelict traps is to determine mortality rates within the traps. As indicated in Appendix C, Table VII, these mortality rates range from 7 to 100 percent. For example, in the case of the Dungeness crab fisheries in the Fraser River Estuary, British Columbia, Breen (1987) estimated that derelict crab pots are responsible for approximately 7 percent of the total catch. The rate of trap loss is also an important factor. Estimates for annual trap loss in the Gulf of Mexico blue crab fishery range from 20 to 100 percent, with higher losses after hurricanes or other severe storms (Guillory et al., 2001). In the Alaskan crab fisheries, during the 1980s, pot losses are thought to have been on the order of 20,000 per year and, even with fewer pots under current limits, pot losses are thought to be about 5,000 per year (Stevens et al., 2000). Although most U.S. trap, cage, and pot fisheries require that pots be equipped with rot cord—sections of twine that compromise the integrity of the pot once they biodegrade—Barnard (2008) determined that the mean failure rate for 30-thread cotton twine is 77–89 days; thus, even properly equipped traps could continue to ghost fish for an extended period. Stevens et al. (2000) reported that one rot cord–equipped ghost pot off Kodiak, Alaska, held 125 crabs. Moreover, there is some anecdotal evidence that compliance with rot cord requirements is incomplete, and that encrusting organisms, such as anemones, can overgrow escape doors and can keep doors functionally closed for years. Derelict pots can also get turned in a way that prevents proper functioning of the escape panels. Ghost fishing losses to hook-and-line gear are poorly documented, but could be substantial for longline gear (National Research Council, 1999).

Other Ecological Impacts

Several studies have suggested that marine debris can act as a transport and dispersal vector for a range of encrusting or clinging species (Winston et al., 1997; Barnes, 2002; Lewis et al., 2005). In a study that confirms this possibility, Zabin et al. (2004) documented the presence of a nonnative sea anemone transported to NWHI on DFG.

Marine debris may also act as an "ecological trap," which affects spatial distributions and migratory patterns or makes large segments of the pelagic community vulnerable to capture. A recent study found that tuna associated with artificial fish aggregating devices (FADs) are less healthy than unassociated tuna (Hallier and Gaertner, 2008). Derelict FADs and other debris assemblages that routinely form along convergence zones or fronts (Pichel et al., 2007) may act as ecological traps by eliciting habitat selection behaviors that are not associated with feeding benefits. Schlaepfer et al. (2002) suggested that the ecological trap effect could be particularly severe when the affected population size is already small (see Chapter 4 for further discussion).

Another impact of marine debris is damage to coral reefs and other benthic communities through entanglement and abrasion. The extent of the damage is dependent on the nature (i.e., size, prevalence, composition) of the debris and the fragility and resilience of the affected environment. In NWHI, Donohue et al. (2001) showed that DFG was responsible for damage to benthic coral reef habitat. Given the continual input of DFG into upcurrent areas (Boland et al., 2006; Dameron et al., 2007), marine debris poses a significant and persistent threat for the reefs of NWHI. In the Florida Keys National Marine Sanctuary, Chiappone et al. (2005) found that hook-and-line fishing gear was the most common type of debris on the reef but noted that the biological impacts were minor, adversely affecting 0.2 percent of the species present.

Socioeconomic Impacts

Marine debris can also reduce direct and indirect socioeconomic benefits (use values, option values, and nonuse values) or increase direct or indirect costs (National Research Council, 2004). Direct benefits include the value of commercial, sport, subsistence, and other cultural harvests; marine transportation; and the benefits that beachgoers, boaters, and divers derive from recreating at the seashore and on marine waters (Smith and Palmquist, 1994; Kaoru et al., 1995; Kirkley and McConnell, 1997; Smith et al., 1997). The following are ways in which marine debris can reduce direct socioeconomic benefits:

- sustainable harvests or catch-per-unit-effort of valued fish and shellfish due to ghost fishing (Kirkley and McConnell, 1997; National Research Council, 1999);
- actual and contingent benefits of coastal recreation due to the presence of litter and other marine debris, including hazardous materials that present human health dangers; and
- net benefits for commercial and recreational boaters from fouling of

propellers and jet intakes as well as damage to hulls (Kirkley and McConnell, 1997).

In addition to direct losses in economic well-being, marine debris can contribute to adverse local economic impacts when beachgoers forego trips to impaired beaches in favor of other recreation opportunities. Ofiara and Brown (1999) and Swanson et al. (1991) estimated that New Jersey lost between $379 million and $3.6 billion in tourism and other revenue as a result of debris washing ashore in 1988. In a South African study, Ballance et al. (2000) estimated that "more than 10 large items per meter of beach would deter 40 percent of foreign tourists, and 60 percent of domestic tourists interviewed, from returning to Cape Town. The impact of this on the regional economy could be a loss of billions of rands each year."

An example of the costs of marine debris removal is presented in an analysis of DFG removal activities by Natural Resources Consultants, Inc. (2007):

> Costs of derelict net survey and removal totaled $4,960 per acre of net removed. Costs of survey and removal of derelict pots/traps totaled $193 per pot/trap. Directly measurable monetized benefits of derelict fishing gear removal were based on the commercial ex-vessel value of species saved from mortality over a one-year period for derelict pots/traps, totaling $248 per pot/trap and a ten-year period for derelict nets, totaling $6,285 per net.

Option benefits reflect the value that individuals derive from reserving the opportunity to engage in coastal recreation or to benefit from coastal amenity services, such as viewing wildlife, at some future time (Bishop, 1982; Freeman, 1984). Awareness of a growing marine debris problem could reduce option value by reducing the probability that individuals will travel to the seashore to recreate or by reducing the benefit they expect to derive from future beach recreation.

Nonuse or vicarious benefits are those obtained from knowledge of the existence of desirable coastal environments, the value derived from being able to bequest unimpaired resources to future generations, the altruistic benefits of preserving attractive coastal resources for other users, and the value associated with the belief that maintaining a litter-free coast and ocean is intrinsically desirable (Brown and Goldstein, 1984; Walsh et al., 1984).

Socioeconomic studies not only help define impacts, but they also can assist with mitigation by improving the understanding of the actions that lead to debris generation. Human behavior is the ultimate cause of marine debris, and the factors that lead to marine debris generation must ultimately be understood and addressed to achieve prevention. At least one study has addressed some of the social aspects of marine debris. Santos et al. (2005b) explored the generation of marine debris on beaches in Brazil and found that tourism was the main source of marine debris; debris levels were correlated with visitor density, and daily litter input to the beach was significantly higher in the regions frequented by people with lower annual income and literacy. Further studies elucidating the role of education level in environmental awareness and human behavior relative to marine debris generation will be needed to ensure a successful long-term solution.

Finding: Despite measures to prevent and reduce marine debris, evidence shows that the problem continues and will likely get worse. This indicates that current measures for preventing and reducing marine debris are inadequate.

Recommendation: Both the United States and the international maritime community should adopt a new approach to prevent and reduce marine debris with more rigorous measures based on a goal of zero discharge of waste into the marine environment.

Finding: While a great deal has been learned about marine debris, there are still many gaps in the understanding of marine debris sources, abundance, fates, and impacts. These gaps in knowledge hinder the ability to prioritize mitigation efforts and to assess the effectiveness of measures that have been implemented.

Recommendation: Additional studies are needed to assess the effectiveness of measures to prevent and reduce marine debris and to provide useful guidance to managers and decision makers for debris mitigation. In particular, the Interagency Marine Debris Coordinating Committee (IMDCC) should sponsor and facilitate research in the following areas:

- Abundance and fluxes: Additional longitudinal marine debris monitoring surveys are needed, particularly for benthic and pelagic debris and for debris fluxes, to identify regional differences and trends in the prevalence, distribution, makeup, and fate of debris. Surveys should pay attention to microplastics as well as macro debris. Survey designs should allow for and encourage comparisons of obtained data.
- Ecological impacts: Future studies on ecological impacts should be designed to provide quantitative estimates of the impact of marine

debris on affected populations and ecosystems and, in addition, have a broad taxonomic focus.
- Socioeconomic impacts: Additional studies should be conducted on the socioeconomics of marine debris, particularly in surveys that explore the human social and behavioral aspects of marine debris generation.

If these studies are to be useful for management, it will be crucial that they be designed in a rigorous manner. Long-term monitoring that allows for comparison of data and evaluative metrics is also important. The elements of a well-designed marine debris survey program are described in the next section.

EFFECTIVE MONITORING AND RESEARCH

The committee's review of the record of marine debris monitoring and research activities found few studies, with some noted exceptions (e.g., Ryan and Moloney, 1993; Henderson, 2001; Barnes, 2005; Sheavly, 2007), that were useful as a reliable indicator of change in marine debris in response to regulatory or other mitigating activities. A large number of studies have been conducted that document aspects of the marine debris problem and the many efforts to manage it, but these are mostly descriptive, anecdotal, and temporal. As discussed in the previous sections, the available body of information documents the complexity of the marine debris problem but does not reliably track the changes in those problems over time and provides little functional insight into the factors controlling and contributing to the marine debris problem. Many studies that purportedly address the effectiveness of MARPOL Annex V address it a posteriori and are unable to link changes in debris definitively to regulatory actions versus other factors. To effectively address marine debris, its scope, sources, causes, and effects, as well as spatial and temporal variability, need to be understood. Mechanisms for objective evaluation must be available to judge the environmental, economic, social, and cultural efficacy of management and mitigation measures. There is a lack of metrics for evaluating the effectiveness of measures implemented to prevent and reduce marine debris, including the effectiveness of specific regulatory and management actions (e.g., education, enforcement actions). Scientifically rigorous monitoring, assessment, and evaluation programs are necessary to understand and address the problem.

Definitions of monitoring, assessment, and evaluation can vary; however, the committee treats them as research activities and advocates conducting them within a structure of systematic, rigorous information collection. In exploring monitoring and research, the importance of research planning and prioritization and the considerations in designing rigorous monitoring and assessment programs, including emerging issues and technologies, are discussed.

Research Planning and Prioritization

Research planning and prioritization need to be driven by a national strategic plan that identifies objectives related to the prevention, mitigation, and remediation of marine debris while remaining nimble enough to address novel issues as they emerge. As described earlier, more research and monitoring are needed to better understand the nature, prevalence, and impacts of marine debris, as well as the effectiveness of measures to address the marine debris problem. Strategic planning and prioritization could also assist in identifying opportunities for incorporating additional technologies or collaborating with other disciplines to maximize the value of marine debris research, given limited resources. Researchers could take better advantage of technologies such as remote sensing and Internet data management and sharing. Data and information gathered from marine debris monitoring, assessment, and research activities could also be incorporated into the emerging national and international efforts to develop an Integrated Ocean Observing System. There are also opportunities for researchers to add onto existing research programs, particularly other longitudinal oceanographic monitoring efforts. Research into the distribution and prevalence of microplastics may be particularly well suited for piggybacking onto existing research programs. Microplastic marine debris has been, and could continue to be, effectively surveyed during ongoing longitudinal plankton surveys run by California Cooperative Oceanic Fisheries Investigations (CalCOFI), the Ecosystems and Fisheries–Oceanography Coordinated Investigations (EcoFOCI), and others. Similarly, sampling of microplastics and other marine debris could be included through additional instrumentation as part of buoys deployed under the Integrated Ocean Observing System. This type of leveraging will add value to and help ensure the long-term sustainability of longitudinal oceanographic monitoring programs.

Finding: A diversity of research on marine debris is conducted, some of it funded by the U.S. federal government, primarily NOAA. However, there is no overall needs assessment available to guide this research. As a result, research completed is rarely integrated at the regional, national, international, or even local levels. Therefore, there is little opportunity for expanding the understanding of marine debris by fitting these individual research activities into a congruous whole.

Recommendation: An information needs assessment should be conducted at the national level by IMDCC with input from stakeholders. A detailed national marine debris research priorities plan should be developed from the results. This research plan should direct future federal funding of a suite of marine debris studies that, when taken together, will provide a comprehensive characterization of the marine debris problem. Such research can serve to inform policy and mitigation actions.

Design of Monitoring, Assessment, and Evaluation Programs

Effective monitoring, assessment, and evaluation programs are also crucial to providing useful information to assess and improve measures to prevent and reduce marine debris. Thoughtful and scholarly analysis of marine debris monitoring, assessment, and evaluation efforts have been completed and many aspects of these works remain relevant (e.g., Ribic, 1990; Ribic et al., 1992). The use of this substantive body of work applied with the benefit of rapidly emerging technologies and advanced analytical methods to current conditions is urgently needed. These efforts must also meet the needs of those seeking to both prevent and mitigate the impacts of marine debris and those needs must be clearly identified.

The characteristics of robust monitoring, assessment, and evaluation activities are relatively straightforward in composition and have been promoted and revisited repeatedly (e.g., Ribic et al., 1992; Lovett et al., 2007). While marine debris presents in a myriad of forms and sizes, standard experimental design principles are applicable and should be employed routinely in future surveys.

As described earlier, NMDMP is an example of a well-designed and scientifically rigorous program for monitoring changes in the composition and prevalence of shoreline marine debris on U.S. coasts. Because a standard sampling protocol was maintained through time and across regions and because sample sites were drawn from a stratified random sample of coast sections, data generated by NMDMP are suitable for a scientifically valid analysis of trends in debris prevalence, across regions and through time. In addition, the data include information on specific indicator items that may be suitable for assessing the effectiveness of targeted source reduction programs. However, the NMDMP survey was conceived as a five-year program and was completed in 2006; there are currently no other ongoing, long-term monitoring programs of this nature. Moreover, NMDMP was designed to provide information about general regional trends and may not be suitable for fine-scale assessments of local trends in deposition or influx from waterways and storm drains.

The United Nations Environment Programme is currently engaged in a global initiative on marine litter with plans to develop targeted regional activities to monitor marine debris, among other activities. This initiative has noted the importance of standardizing monitoring protocols to allow for regional comparisons; it plans to develop "substantive guidelines and recommended policies on harmonizing monitoring systems of marine litter" (United Nations Environment Programme, 2008). This initiative provides an opportunity for the United States and other nations to coordinate their monitoring efforts to ensure comparability.

Finding: Well-designed and statistically rigorous longitudinal marine debris monitoring programs are needed at a variety of spatial and temporal scales. However, standardization of protocols is necessary to ensure that the results of various surveys are comparable.

Recommendation: Long-term marine debris monitoring programs should be established by IMDCC (for the United States) and appropriate international organizations such as the United Nations Environment Programme (for global monitoring). These programs should allow for statistically valid analysis of marine debris quantities and trends as a metric of the effectiveness of measures to prevent and reduce marine debris. To the extent practical, these programs should adopt a suite of common design characteristics and protocols to facilitate cross comparisons and meta-analyses.

Remote sensing using satellites, planes, remotely operated vehicles, and other devices represents a promising technology for assessing the nature and extent of marine debris and enhancing understanding and mitigation because of its ability to systematically observe and measure large or otherwise inaccessible areas of the ocean. For example, airborne synthetic aperture radar is a type of remote sensing instrument that has the potential to identify, map, and guide the removal of plastic debris at sea at very fine scales, particularly for large debris items such as DFG. While synthetic aperture radar is a fairly new technology and not readily available, there are also a variety of similar existing tools available to researchers (e.g., remote sensing instruments mounted on U.S. Coast Guard aircraft) that could be used in studies and efforts to mitigate marine debris.

Remote sensing, in combination with ocean circulation models, shows particular promise in identifying areas of debris accumulation in the open ocean for targeted remediation efforts (Kubota, 1994; Ingraham and Ebbesmeyer, 2001; Polovina et al., 2001; Bograd et al., 2004; Kubota et al., 2005; Pichel et al., 2007). An example of a public–private partnership that is using remote sensing data from both satellites and aerial surveys to mitigate marine debris is the "GhostNet Project" (Airborne Technologies Incorporated, Wasilla, Alaska), which successfully located over 2,000 debris items in the open ocean (Pichel et al., 2007). Using remote sensing data for mitigation can be challenging given the difficulty in verifying marine debris locations with sufficient precision and within acceptable timeframes and costs to direct removal efforts (e.g., ship-based recovery of marine debris), particularly given the mobile nature of floating marine debris. Although the recovery of marine debris via dedicated ships can be costly (e.g., Donohue [2005] reported $30,000 per ton), targeting these efforts by identifying high-density debris areas using remote sensing could significantly minimize operational costs.

Remote sensing has also proven helpful in understanding marine debris impacts. For example, remote sensing has been used to detect phytoplankton blooms caused by floating plastic, which provides an artificial substratum (Mato et al., 2001; Barnes, 2002). Morishige et al. (2007) used remote sensing to show that marine debris deposition in NWHI is influenced by the El Niño/La Niña phenomenon, while Donohue and Foley (2007) used remote sensing to show that Hawaiian monk seal entanglement is greater in El Niño years.

Marine Debris Information Clearinghouse

There is a significant opportunity with regard to the dynamic use of the Internet and other emerging technologies for practitioners whose actions may directly influence marine debris generation and mitigation. The marine debris "clearinghouse," as called for in the recent U.S. Marine Debris Research, Prevention, and Reduction Act (33 U.S.C. § 1951 et seq.), has the potential to contribute to this purpose if crafted with consideration of the multiple users that influence marine debris sources and solutions. Data management is an important component of such a clearinghouse site both to make marine debris data widely available and to promote standardized marine debris data collection protocols. There are many good examples of protocols for establishing a data archive meta-database system and for providing ready access. The Marine Conservation Alliance Foundation's Google™ map interface (Marine Conservation Alliance Foundation, 2008) is an example of a user-friendly way to layer information on the distribution of debris and the conduct of cleanup activities.

Finding: The value of data stored in the marine debris information clearinghouse, mandated by the Marine Debris Research, Prevention, and Reduction Act, will depend on how well it is standardized, how well it is integrated into a meta-database, and whether it is readily accessible to researchers and the public.

Recommendation: The marine debris information clearinghouse should be given high priority. It should be housed and maintained by NOAA but available to the public and researchers at large. Data generated by federally funded research should be submitted to this clearinghouse in a timely manner.

CONCLUSION

The following finding and recommendation express overarching concepts discussed in the previous findings and recommendations in Chapter 2.

Overarching Finding: Although there is clear evidence that marine debris is a problem, there has not been a coordinated or targeted effort to thoroughly document and understand its sources, fates, and impacts. This confounds the ability to prioritize mitigation efforts and to assess the effectiveness of measures that have been implemented.

Overarching Recommendation: IMDCC should, through planning and prioritization, target research to understand the sources, fates, and impacts of marine debris. It should support the establishment of scalable and statistically rigorous protocols that allow monitoring at a variety of temporal and spatial scales. These protocols should contain evaluative metrics that allow an assessment of progress in marine debris mitigation. The United States, through leadership in the international arena, should provide technical assistance and support for the establishment of additional monitoring and research programs worldwide.

3

Measures to Prevent and Reduce Marine Debris and Its Impacts

A variety of laws, regulations, and nonregulatory measures can be applied to prevent or limit impacts of the disposal of garbage into the oceans. There are, however, no comprehensive programs designed to assess the amount and impacts of debris that is already in or will make its way into the oceans, or to remediate and remove that debris. This chapter reviews and identifies gaps in the existing international legal and regulatory framework, including port reception facilities. It then discusses and identifies gaps in U.S. domestic laws that are most relevant to prevention and reduction of marine debris from land as well as ocean-based sources. The chapter also addresses U.S. implementation of these regulations related to leadership and coordination, integrated solid waste management, waste minimization and source reduction, enforcement and compliance activities, and mitigation and removal programs.

INTERNATIONAL LEGAL AND REGULATORY FRAMEWORK

There are two primary international conventions that address garbage pollution in the oceans: the International Convention for the Prevention of Pollution from Ships, 1973, as modified by the Protocol of 1978 Annex V and the Convention on the Prevention of Marine Pollution by Dumping of Wastes and Other Matter, 1972, and the 1996 Protocol to the Convention. The overarching framework for these international conventions is set in the United Nations Convention on the Law of the Sea.

United Nations Convention on the Law of the Sea

The basic principles of international ocean law are set forth in the 1982 United Nations Convention on the Law of the Sea. This comprehensive treaty, which entered into force in 1994, describes the rights and responsibilities of nations to conduct and control activities in and affecting the oceans. Although the United States has not ratified the Convention on the Law of the Sea, the Executive Branch has submitted it to the U.S. Senate for advice and consent with a recommendation that it be ratified and that the United States considers most of its provisions to reflect binding customary international law (Van Dyke, 2008). The Convention sets out a number of duties that are relevant to the global marine debris problem (Box 3.1). These duties oblige nations to use their authority and

---

BOX 3.1
Marine Debris Pollution and the United Nations
Convention on the Law of the Sea

Provisions of the United Nations Convention on the Law of the Sea that require nations to combat marine debris include the following:

Article 1: For the purposes of this Convention: …(4) "pollution of the marine environment" means the introduction by man, directly or indirectly, of substances or energy into the marine environment, including estuaries, which results or is likely to result in such deleterious effects as harm to living resources and marine life, hazards to human health, hindrance to marine activities, including fishing and other legitimate uses of the sea, impairment of quality for use of sea water and reduction of amenities.

Article 192: States have the obligation to protect and preserve the marine environment.

Article 194: (1) States shall take … all measures necessary to prevent, reduce, and control pollution of the marine environment from any source. . . . (5) The measures taken in accordance with this part shall include those necessary to protect and preserve rare or fragile ecosystems as well as the habitats of depleted, threatened or endangered species and other forms of marine life.

Article 197: States shall cooperate on a global basis, and as appropriate, on a regional basis, directly or through competent international organizations, in formulating and elaborating international rules, standards and recommended practices and procedures . . . for the protection and preservation of the marine environment, taking into account characteristic regional features.

Article 207: (1) States shall adopt laws and regulations to prevent, reduce and control pollution of the marine environment *from land-based sources*, including

continued

BOX 3.1 Continued

rivers, estuaries, pipelines and outfall structures, taking into account internationally agreed rules, standards and recommended practices and procedures. . . . (5) Laws, regulations, measures, rules, standards and recommended practices and procedures . . . shall include those designed to minimize, to the fullest extent possible, the release of toxic, harmful or noxious substances, especially those which are *persistent*, into the marine environment.

Article 210: (1) States shall adopt laws and regulations to prevent, reduce and control pollution of the marine environment by dumping. (2) States shall take other measures as necessary to prevent, reduce and control such pollution . . . (4) States acting especially through competent international organizations or diplomatic conference, shall endeavor to establish global and regional rules, standards and recommended practices and procedures to prevent, reduce and control such pollution . . . (6) National laws, regulations and measures shall be no less effective in preventing, reducing and controlling such pollution than global rules and standards.

Article 211: (1) States, acting through the competent international organization or general diplomatic conference, shall establish international rules and standards to prevent, reduce and control pollution of the marine environment *from vessels . . . Such rules and standards shall . . . be re-examined from time to time as necessary.* (2) States shall adopt laws and regulations for the prevention, reduction and control of pollution of the marine environment from vessels flying their flag or of their registry . . . (3) States which establish particular requirements . . . as a condition for entry of foreign vessels into their ports or internal waters . . . shall . . . communicate them to the competent international organization.

Article 216: (1) Laws and regulations adopted in accordance with this Convention and applicable international rules and standards established through competent international organizations or diplomatic conference for the prevention, reduction and control of pollution of the marine environment by dumping shall be enforced: (a) by the coastal State with regard to dumping within its territorial sea or its exclusive economic zone or onto its continental shelf; (b) by the flag State with regard to vessels flying its flag or vessels or aircraft of its registry; (c) by any state with regard to acts of loading of wastes or other matter occurring within its territory or at its off-shore terminals.

Article 217: States shall ensure compliance by vessels flying their flag or of their registry with applicable international rules and standards . . .

Article 218: (1) When a vessel is voluntarily in a port . . . of a state, that state may undertake investigations and, where the evidence so warrants, institute proceedings in respect of any discharge from that vessel outside the internal waters, territorial sea or exclusive economic zone of that state in violation of applicable international rules and standards established through the competent international organization or general diplomatic conference.

SOURCE: United Nations Convention on the Law of the Sea of 1982 (emphasis added).

jurisdiction to prevent degradation of the marine environment, including prevention of land- and ocean-based discharges of marine debris. The Convention encourages nations to act through international bodies, such as the International Maritime Organization (IMO), but makes it clear that nations have a continuing legal duty to exercise the full extent of their authorities over activities on land and at sea to supplement internationally agreed measures.

The Convention on the Law of the Sea refers to national regulations to prevent marine pollution, as well as standards that are adopted through "competent international organizations" (United Nations Convention on the Law of the Sea of 1982, Article 61) for pollution from vessels. With regard to shipping and marine debris, IMO is the responsible body.

International Convention for the Prevention of Pollution from Ships, 1973, as modified by the Protocol of 1978

IMO, a specialized agency of the United Nations, was created in 1948 to establish consistent international regulation of the maritime industry. Membership in IMO includes 167 nations; several nongovernmental and intergovernmental organizations also participate in a consultative status (International Maritime Organization, 2008a). Through its specialized committees and subcommittees, the IMO Assembly has created a comprehensive body of international conventions and supporting annexes to govern international maritime commerce.

Although IMO initially focused on developing regulations to promote safety, vessel accidents that resulted in significant pollution events led to IMO initiatives to include prevention and management of pollution associated with accidents and normal operations. The most significant of these initiatives is the International Convention for the Prevention of Pollution from Ships, 1973, as modified by the Protocol of 1978 (MARPOL). In its current form, MARPOL contains six operational annexes. These annexes address prevention of pollution by oil (Annex I), control of pollution by noxious liquid substances in bulk (Annex II), prevention of pollution by harmful substances carried by sea in packaged form (Annex III), prevention of pollution by sewage from ships (Annex IV), prevention of pollution by garbage from ships (Annex V), and prevention of air pollution from ships (Annex VI). Parties wishing to ratify MARPOL must ratify Annexes I and II. Ratification of the other annexes, including Annex V, is optional. When a nation agrees to become a "party" to an agreement, such as MARPOL Annex V, it is required to adopt domestic legislation to ensure implementation of the treaty requirements. In the United States, ratification requires the advice and consent of the Senate, enactment of enabling legislation, and appropriation of requisite funding. The United

States ratified MARPOL Annex V in 1987. Currently, 134 nations representing nearly 97 percent of the world's tonnage are parties to MARPOL Annex V, which entered into force on December 31, 1988 (International Maritime Organization, 2008b).

MARPOL Annex V: Prevention of Pollution by Garbage from Ships

MARPOL Annex V seeks to eliminate or reduce the disposal of garbage from ships by specifying the conditions under which different types of garbage may be discharged. MARPOL Annex V prohibits the at-sea disposal of plastics of any kind and tightly restricts other discharges in coastal waters and designated "special areas."

MARPOL Annex V has been amended twice since it entered into force on December 31, 1988. In 1994, an amendment on port state control provisions was added, which establishes the framework for parties to ensure and promote compliance with the provisions of MARPOL Annex V through national inspection and enforcement programs applicable to vessels and shoreside facilities. In the United States, these programs apply to U.S. flag vessels located anywhere in the world, to foreign flag vessels in the territorial waters of the United States and calling in U.S. ports, and to shoreside facilities that are required to provide adequate reception facilities to vessels berthed at those facilities. In 1995, an amendment was added requiring garbage management plans and record books for all ships 400 gross tons and above and those certified to carry 15 persons or more, and placarding for all ships 12.192 meters (40 feet) or more in length. The garbage management plans and record books are discussed in further detail below.

Under MARPOL Annex V, garbage is defined as "all kinds of victual, domestic, and operational waste excluding fresh fish and parts thereof, generated during the normal operation of the ship and liable to be disposed of continuously or periodically except those substances which are defined or listed in other Annexes to the present Convention" (International Maritime Organization, 2006d). Under the exception, it is clear that if components of the materials to be discharged are covered by more specific and stringent provisions of other MARPOL annexes, the more stringent provisions prevail. A more complicated question that is currently under review by parties to MARPOL (the review is being conducted by IMO's Marine Environment Protection Committee; see section below for further discussion) is how to treat garbage that contains marine pollutants or harmful and hazardous substances not specifically covered by other MARPOL Annexes (International Maritime Organization, 2007). MARPOL requirements apply to all covered ships at all times, with the exception from certain requirements for vessel emergencies that require a discharge to secure the safety of the ship and human life or that result from actions taken to secure the safety of the ship and human life. The accidental loss of synthetic fishing nets is also exempt under MARPOL Annex V, provided that reasonable precautions have been taken to prevent the loss (see detailed discussion in Chapter 4). Table 3.1 outlines the garbage management framework established by MARPOL Annex V.

In general, MARPOL Annex V establishes a "distance from land" framework for permissible dumping of garbage with more strict prohibitions in special areas. These distances (3, 12, and 25 nautical miles) are based primarily on historical definitions of state, territorial seas, and

TABLE 3.1 Summary of Garbage Discharge Restrictions for Vessels (modified from International Maritime Organization, 2006b)

Garbage Type	All Ships, Except Platforms		Offshore Platforms[a]
	Outside Special Areas	Inside Special Areas	
Plastics—includes synthetic ropes and fishing nets and plastic garbage bags	Disposal prohibited	Disposal prohibited	Disposal prohibited
Floating dunnage, lining, and packing materials	>25 nautical miles offshore	Disposal prohibited	Disposal prohibited
Cargo residues, paper, rags, glass, metal, bottles, ash and clinkers, crockery, and similar refuse	>12 nautical miles offshore	Disposal prohibited	Disposal prohibited
All other garbage, including paper, rags, and glass, comminuted or ground[b]	>3 nautical miles offshore	Disposal prohibited	Disposal prohibited
Food waste not comminuted or ground	>12 nautical miles offshore	>12 nautical miles offshore	Disposal prohibited
Food waste comminuted or ground[b]	>3 nautical miles offshore	>12 nautical miles offshore[c]	>12 nautical miles offshore

[a]Offshore platforms and associated ships include all fixed or floating platforms engaged in exploration or exploitation of sea-bed mineral resources and all ships within 500 m of such platforms.
[b]Comminuted or ground garbage must be able to pass through a screen with mesh size no larger than 25 mm.
[c]For the Wider Caribbean Region, disposal is allowed at greater than 3 nautical miles offshore.

international waters rather than ecosystem considerations. However, the more stringent restrictions within designated special areas reflect broader environmental concerns. A key to the implementation of MARPOL Annex V is the requirement that parties provide adequate garbage reception facilities for ships calling at their ports and terminals. Despite the permissibility of at-sea discharges in compliance with MARPOL Annex V, it is environmentally prudent for vessels to discharge their garbage ashore where it can be handled by more sophisticated (in developed nations) solid waste management systems, which often include recycling and waste treatment programs.

MARPOL Annex V ships are required to maintain a garbage management plan, which sets out written procedures for the collection, storage, processing, and disposal of all types of garbage generated on the vessel, including operating guidelines for solid waste management equipment installed aboard the vessel. The garbage management plan and record book provide a written record of all garbage discharges and incineration at sea, including the date, time, position of the vessel, and description of the type of garbage discharged or incinerated. In addition, the garbage record book must include records of accidental and willful discharges of garbage that are not compliant with the provisions of MARPOL Annex V, along with a description of the circumstances and reasons for the discharge (e.g., emergency situations). An appendix to MARPOL Annex V contains a guideline and template for the garbage record book and details on the nature of entries to be recorded.

As is the case with many other international conventions, IMO has agreed to a nonmandatory set of guidelines which may be used by nations in developing legislation for the implementation of MARPOL Annex V. *Guidelines for the Implementation of Annex V of MARPOL* (International Maritime Organization, 2006b) adds details to the provisions of MARPOL Annex V. This document also includes three appendices that address reporting of alleged inadequacies of port reception facilities, specifications for shipboard incinerators, and guidance for the development of garbage management plans. For example, the International Maritime Organization (2006b) details solid waste management options, such as waste minimization and onboard garbage processing, to assist in MARPOL Annex V compliance. Figure 3.1 shows the possible options for shipboard handling and disposal of garbage, from collection through disposal.

As previously noted, MARPOL provides for designation of special areas that provide a higher level of protection than other areas; as indicated in Table 3.1, the only authorized discharge of garbage into a special area is food waste, except under emergency circumstances. Eight special areas have been designated by MARPOL Annex V: the Baltic Sea, the Mediterranean Sea, the Black Sea, the Red Sea, the North Sea, the Wider

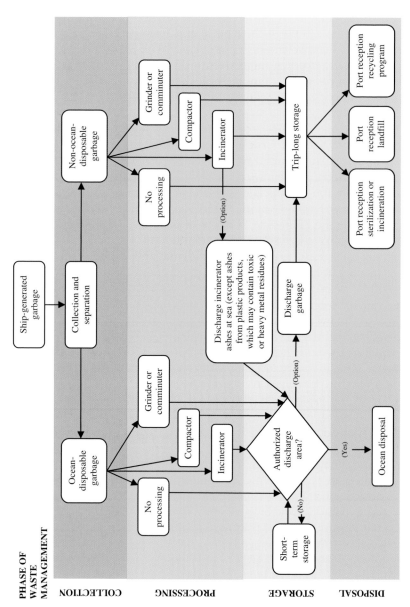

PHASE OF WASTE MANAGEMENT

COLLECTION

PROCESSING

STORAGE

DISPOSAL

FIGURE 3.1 Options for shipboard handling and disposal of garbage (modified from International Maritime Organization, 2006b).

Caribbean Region, the Antarctic, and the (Arabian) Gulfs area. However, only three of these special areas (the North Sea, the Baltic Sea, and the Antarctic region) have been in effect for several years. The Gulfs special area went into effect in August 2008, and the Mediterranean special area will go into effect in May 2009, leaving the Black Sea, the Red Sea, and the Wider Caribbean Region without the protections from marine debris deemed necessary by their initial special area designation. The single most significant obstacle to implementation of marine debris protection programs in these areas is the lack of certification of adequate reception facilities. There are no specific incentives or technical assistance to provide adequate shoreside waste disposal facilities, and, without having facilities in place, ships are not required to provide the added level of protection to these special areas. IMO could do more to assist parties bordering on special areas to meet their obligations. However, the general failure of the special area provisions to actually provide extra protection to these sensitive areas indicates that a new approach is needed.

Convention on the Prevention of Marine Pollution by Dumping of Wastes and Other Matter, 1972, and the 1996 Protocol to the Convention

The Convention on the Prevention of Marine Pollution by Dumping of Wastes and Other Matter (commonly referred to as the London Convention) was agreed to in 1972 and entered into force in 1975. The Convention focuses on preventing the dumping of wastes and other materials into the sea. The Protocol to the London Convention (commonly referred to as the London Protocol), agreed to in 1996 and entered into force on March 24, 2006, updates the Convention. It is anticipated that the Protocol will supplant the Convention in its entirety; the United States is currently in the process of ratifying the Protocol. Therefore, this summary will focus on the provisions and requirements established by the Protocol.

Under the Protocol, dumping is defined as the following:

1. any deliberate disposal into the sea of wastes or other matter from vessels, aircraft, platforms or other man-made structures at sea;
2. any deliberate disposal into the sea of vessels, aircraft, platforms or other man-made structures at sea;
3. any storage of wastes or other matter in the seabed and the subsoil thereof from vessels, aircraft, platforms or other man-made structures at sea; and
4. any abandonment or toppling at site of platforms or other man-made structures at sea, for the sole purpose of deliberate disposal (1996 Protocol to the Convention, Article 1).

Perhaps equally important is that dumping does *not* include

> the disposal into the sea of wastes or other matter incidental to, or derived from the normal operations of vessels, aircraft, platforms, or other man-made structures at sea and their equipment, other than wastes or other matter transported . . . for the purpose of disposal of such matter or derived from the treatment of such wastes . . . (1996 Protocol to the Convention, Article 1).

Put simply, discharges from vessels, aircraft, platforms, or other man-made structures at sea are not considered dumping if they are wastes generated during "normal operations"; however, they are considered by the Convention and the Protocol if the discharged materials were transported for the express purpose of disposal at sea. Other provisions of the Protocol prohibit the at-sea incineration of wastes covered under the Protocol and also prohibit the export of wastes to other countries for subsequent dumping or incineration at sea.

A key difference between the Convention and the amended Protocol is that where the Convention allowed dumping unless specifically prohibited (a so-called "black list" approach), under the Protocol, at-sea dumping is prohibited unless the material has been specifically included on an approved list (a "reverse list" or "white list" approach). The Protocol also incorporates a precautionary approach to protecting the marine environment from dumping activities by requiring preventative action to be "taken when there is reason to believe that wastes or other matter introduced into the marine environment are likely to cause harm even when there is no conclusive evidence to prove a causal relation between inputs and their effects" (1996 Protocol to the Convention, Article 3).

Annex I of the Protocol lists wastes that may be dumped pursuant to a permit. This white list includes dredged material; sewage sludge; fish waste; vessels and platforms or other man-made structures; inert, inorganic geological material; organic material of natural origin; and bulky items primarily comprising iron, steel, concrete, and other minimally harmful materials. Annex II of the Protocol establishes procedures for assessment of wastes that are being considered for dumping and includes provisions related to solid waste prevention, solid waste management, dump-site selection, assessment of potential impacts of solid waste management options, compliance and monitoring programs, and criteria for issuing permits and establishing appropriate permit conditions specific to a particular material. Parties to the Protocol must issue permits (when deemed acceptable) for materials that are loaded in their territory, regardless of country of registry, and to vessels or aircraft registered in their territory if such loading of covered materials occurs in the territory of a nation not a party to the Protocol.

In 2006, generic and waste-specific guidelines for assessment of wastes which may be dumped under permitted conditions were updated and published by IMO as *Guidelines on the Convention on the Prevention of Marine Pollution by Dumping of Wastes and Other Matter, 1972* (International Maritime Organization, 2006d). These guidelines provide specific criteria and evaluation processes for the assessment of the wastes that are listed in Annex I to the Protocol as permissible for dumping.

GAPS IN THE INTERNATIONAL LEGAL AND REGULATORY FRAMEWORK

The U.S. Commission on Ocean Policy (2004) notes that

> The dominant paradigm for governing the oceans [had been] the principle of freedom of the seas, based on the premise that the oceans were infinite and marine resources inexhaustible. . . . This view of the oceans began to change dramatically in the middle of the 20th century, when it became apparent that problems of overfishing and pollution threatened ocean assets that had previously been taken for granted.

This statement reflects the growing awareness of fragility of the ocean ecosystems, as well as a cultural and policy shift away from the operational and vessel focus of the early days of IMO toward a more ecosystem-based view that emphasizes the minimization, and ideally elimination, of discharges of garbage and other debris into the marine environment. This shift is demonstrated most clearly with the move from a black list approach to regulating ocean dumping in the London Convention under which material could be dumped unless prohibited to a more precautionary white list approach in the London Protocol, which presumes that material should not be dumped at sea. Although the voluntary guidelines for implementing MARPOL Annex V (International Maritime Organization, 2006b) do include waste minimization and source reduction as significant objectives, the current mandatory framework is constructed on the premise that discharges are permissible as long as they are consistent with applicable conditions and are not expressly prohibited.

Nearly 20 years after MARPOL Annex V was originally adopted, and with advances in ship operating procedures, available technologies, and solid waste management practices, it would be reasonable to consider many of the positive discharge mitigation philosophies embodied in the *Guidelines* (International Maritime Organization, 2006b) for inclusion into the mandatory legal requirements of MARPOL Annex V. Concepts of waste minimization, source reduction, and zero discharge are being successfully employed in many industries, including some segments of the maritime industry (see "Waste Minimization and Source Reduction").

However, adoption and implementation of these approaches by the maritime sector has not been a priority under the current MARPOL framework. While the *Guidelines* (International Maritime Organization, 2006b) provides a good start, a paradigm shift will require a more significant commitment to technology development and development of best practices for shipboard solid waste management, and incorporation of successful practices into MARPOL Annex V regulatory standards. Shipboard garbage management will depend on fleet characteristics such as vessel size, passenger and crew numbers, routes and ports of call, and average voyage length. The National Research Council (NRC) (1995a) identified obstacles to and strategies for compliance with MARPOL Annex V for nine different fleets; the obstacles remain, and the suggested strategies are still appropriate.

Marine Environment Protection Committee Correspondence Group

The United Nations General Assembly asked IMO "to review MARPOL Annex V, in consultation with relevant organizations and bodies, and to assess its effectiveness in addressing sea-based sources of marine debris" (International Maritime Organization, 2006a). In response, the Marine Environment Protection Committee (MEPC), a subcommittee of the IMO, established an intersessional correspondence group. The correspondence group was asked "to develop the framework, method of work, and timetable for a comprehensive review of MARPOL Annex V *Regulations for the Prevention of Pollution by Garbage from Ships* and the associated *Revised Guidelines for the Implementation of MARPOL Annex V*" (International Maritime Organization, 2006a). Specifically, the correspondence group was tasked with the following:

- examination of Annex V and its *Guidelines;*
- consideration of the issues submitted;
- an assessment of trends in sea-based sources of marine debris;
- consideration of relevant work of other bodies; and
- development of necessary amendments to Annex V and its *Guidelines* (International Maritime Organization, 2007).

In an interim report (International Maritime Organization, 2007), the correspondence group indicated that progress has been made on the first four tasks and that it anticipated completion of the last task in July 2008. Issues of relevance to this report, which are being addressed by the correspondence group but for which no final recommendations have been made, include

- creation of a general prohibition on the discharge of garbage;
- integration of waste minimization principles into MARPOL Annex V and the *Guidelines*;
- additional placarding and recordkeeping requirements;
- adequacy of reception facilities;
- management of cargo residues in general and in special areas;
- management of bulk liquid wastes not subject to other MARPOL annexes;
- management of oils used in the ship's galley;
- garbage that may contain harmful residues that are not currently defined as pollutants;
- discharge of floating dunnage and packaging materials;
- discharge of composite materials; and
- mitigation of the loss of fishing gear and the promotion of responsible fishing practices.

The correspondence group has addressed each of these issues through the identification of options that need to be considered prior to finalization of their report, which is expected to include recommendations for future action including amendments to MARPOL Annex V and which will be considered by MEPC in October 2008.

Finding: Under MARPOL Annex V, as currently written, discharges are permitted unless specifically prohibited. This approach does not provide sufficient incentive to encourage innovation and adoption of source reduction and waste minimization measures to prevent garbage pollution in the marine environment.

Recommendation: The U.S. delegation to IMO should, through the ongoing review process, advocate that IMO amend MARPOL Annex V to include a general prohibition on discharge of garbage at sea with limited exceptions based on specific vessel operating scenarios and adequacy of shoreside reception facilities. In addition, the U.S. delegation should request that IMO review the *Guidelines for the Implementation of Annex V of MARPOL* (International Maritime Organization, 2006b) and, where transferrable, amend MARPOL Annex V to include waste minimization and source reduction concepts from the *Guidelines* into mandatory requirements for vessels, such as within garbage management plan requirements. The United States and other parties to MARPOL Annex V should incorporate similar requirements into their domestic regulations for vessels engaged in both international and domestic trade.

Port Reception Facilities

The lack of understanding of the marine waste and vessel stream and the inadequacy of shoreside reception to accept and properly manage vessel waste is a serious impediment to prevention and reduction of marine debris, including derelict fishing gear (DFG). Moreover, the lack of uniform disposal fees, particularly for contaminated wastes, provides a financial incentive for vessel operators to discard at sea when such discards are permitted under MARPOL Annex V or when the probability of detection is deemed low. Providing incentives for landside disposal of ship-generated waste is a practical method for curbing waste discharge at sea. Some examples of port reception financing systems that may promote landside disposal include the following:

- "free of charge" systems, which charge no fees to ships for waste reception, handling, or disposal;
- "non-special fee" systems, in which the cost of reception, handling, and disposal is included in overall port fees or general environmental fees and charged regardless of whether or not waste is offloaded;
- "fixed fee" systems, which are similar to the non-special fee systems, but the waste disposal cost is a separate fixed fee and is paid regardless of whether or not the ship offloads waste; and
- "deposit-refund" systems which charge ships a mandatory waste management fee as a deposit, then refund all or part of this fee to those ships that use the port reception facility services (Olson, 1994; Carpenter and Macgill, 2001; Georgakellos, 2007).

The non-special fee system has been employed in the Baltic region since the late 1990s under the Baltic Strategy of the Helsinki Commission (HELCOM), in an effort to prevent illegal discharges of waste at sea and to provide economic incentives to dispose onshore. The HELCOM Baltic Sea Action Plan, adopted in 2007, extended the existing non-special fee system for ship-generated wastes, but it also recommended that litter caught in fishing nets be covered in the non-special fee system. It further requested that its own members support and seek active cooperation with the neighboring North Sea Region for adoption of a similar non-special fee system for garbage. Additionally, since the Baltic Sea is designated a special area under MARPOL Annex V, and all discharges of garbage at sea are prohibited, HELCOM requires all ships to offload their garbage ashore prior to leaving Baltic ports (Helsinki Commission, 2007).

Even without additional requirements as recommended in this document, ships continue to face shoreside disposal challenges at some berths in countries which have formally communicated the availability of adequate reception facilities.[1] In its previous report, the NRC noted that, in the United States, only "general guidance is provided . . . but there are no technical standards" for port adequacy, leaving wide variations on the interpretation of "adequacy" (National Research Council, 1995a). For example, the U.S. Coast Guard (USCG) Certificate of Adequacy (COA) program bases its certification not on whether the ports actually accept shipborne garbage, but on whether they are *capable* of accepting garbage or can demonstrate that they have service providers on call who can accept the garbage.[2] While it is well understood that such a service is not usually provided free of charge while docking at these berths, vessels, ready and willing to pay for disposal services either directly from the facility or via independent entities, are not always able to secure these services, even from those ports with COAs.

Finding: While parties to MARPOL Annex V are required to ensure adequate port reception facilities, the standards for adequacy are unclear. Throughout the world, ships continue to encounter obstacles to discharging garbage ashore at some ports that have otherwise formally communicated the availability of adequate reception facilities for MARPOL Annex V waste. Although the *Guidelines for the Implementation of Annex V of MARPOL* (International Maritime Organization, 2006b) provides additional guidance, it does not establish minimum standards.

Recommendation: The U.S. delegation to IMO should advocate that MARPOL Annex V be amended to include explicit qualitative and quantitative standards for adequate port reception facilities, and that IMO provide assistance to achieve these standards. Port managers and users should be included in the development of clearer standards. In addition, the U.S. delegation should encourage IMO to incorporate incentives for proper onshore waste disposal in these standards. In the United States, USCG should incorporate these minimum standards into their COA program and should encourage ports to provide incentives to vessel operators for discharging their waste ashore.

---

[1]IMO maintains a Global Integrated Shipping Information System website (free to the public after registration) that contains a Port Reception Facility module, which "contains information on the available port reception facilities for the delivery of the ship-generated waste, as provided by the competent authorities of the IMO member states" (International Maritime Organization, 2008c). It also contains a database of reports of alleged inadequacies.

[2]USCG maintains an online database of U.S. ports and terminals that hold valid MARPOL COAs (U.S. Coast Guard, 2008).

DOMESTIC LEGAL, REGULATORY, AND MANAGEMENT FRAMEWORK

The U.S. domestic implementing legislation for the international conventions described above overlaps with a myriad of other laws that guide the prevention, reduction, and management of marine debris. Federal authorities and responsibilities relevant to marine debris management are spread across several agencies. The lead agencies for marine debris and their primary responsibilities include the following:

- the National Oceanic and Atmospheric Administration (NOAA), whose principal responsibilities include monitoring, research and education, fisheries management, and response and restoration;
- USCG, which is responsible for ship- and port-related issues, including the COA program for port reception facilities and domestic implementation and enforcement of MARPOL Annex V; and
- the Environmental Protection Agency (EPA), which is responsible for regulating and monitoring the environmental impacts of garbage and land-based sources of marine debris, regulation of pollution discharged into coastal and marine waters, and domestic implementation of the London Convention and amended Protocol.

U.S. management of marine debris is further complicated because the aforementioned federal agencies are not directly responsible for providing waste disposal services when garbage and other wastes reach or are generated on land. Although disposal, recycling, incineration, and landfill sites are regulated nationally by EPA, the responsibility for implementation falls to the state and local levels and is organized by local government and service districts that may provide those services directly or enter into contracts with private service providers.

The major domestic laws implementing requirements of MARPOL Annex V and the London Protocol, as well as other U.S. laws most relevant to the prevention, reduction, and management of marine debris, are summarized.

Act to Prevent Pollution from Ships of 1982 and the Marine Plastic Pollution Research and Control Act of 1987

The Act to Prevent Pollution from Ships (APPS) (33 U.S.C. § 1901 et seq.) was adopted in 1980. It was amended by the Marine Plastic Pollution Research and Control Act of 1987 (MPPRCA) (33 U.S.C. § 1901 et seq.) to implement the provisions of MARPOL Annex V. USCG has the primary responsibility under APPS for establishing and enforcing regulations, which require ships to maintain garbage record books and shipboard management plans and to display placards (Figure 3.2) that notify the crew and passengers of the requirements of MARPOL Annex V; USCG also has responsibility to specify the ships to which the regulations apply (33 U.S.C. § 1903(a)). In 1991, the rules were amended to give effect to an IMO amendment to MARPOL Annex V eliminating an exemption for the loss of synthetic material incidental to the repair of fishing nets.

APPS and its implementing regulations generally follow the provisions and implementing regulations of MARPOL Annex V as set out in Table 3.1 (33 C.F.R. §§ 151.51–151.77). The discharge of garbage is prohibited within 3 nautical miles of shore, and the discharge of specified types of materials is prohibited 3–12 and 12–25 nautical miles from shore. The discharge of plastics at sea is prohibited, including synthetic ropes and synthetic fishing gear. All U.S. manned oceangoing vessels that are 26 feet or longer are required to display placards that notify crew of disposal restrictions. Every ship of 400 gross tons or more, manned fixed or floating platforms, and every ship certified to carry more than 15 passengers engaged in international voyages must keep a written record of discharge, disposal, and incineration operations. Oceangoing ships longer than 40 feet documented under U.S. laws and U.S. fixed and floating platforms must have waste management plans. USCG has the authority to conduct inspections of vessel discharge records and logbooks, to respond to

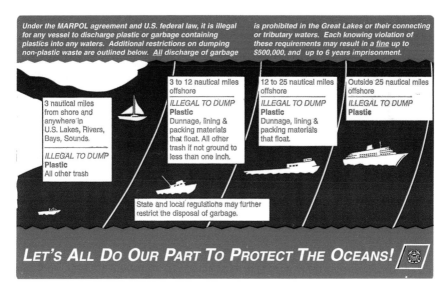

FIGURE 3.2 Example of an APPS placard of garbage discharge restrictions for vessels (used with permission from the U.S. Coast Guard).

reports of illegal discharges, and to impose fines on violators. The provisions of APPS relating to garbage and plastics apply to all U.S. ships operating anywhere in the world, and to foreign flag vessels operating in U.S. navigable waters, the U.S. Exclusive Economic Zone (EEZ), or at a port under the jurisdiction of the United States.

In 1994, USCG did a study of compliance with the 1987 Act and regulations because, despite implementation of MARPOL Annex V, large quantities of plastic continued to wash ashore, obstruct navigation, and entangle marine life. USCG found that many vessels had neither plastic waste nor residue from incineration of plastic wastes onboard and that less than 20 percent of vessels calling at ports on the east and Gulf coasts off-loaded garbage at reception facilities (59 Fed. Reg. 18700 [April 19, 1994]). Because it was improbable that these vessels did not generate plastic waste and plastics continue to be a large component of the marine debris problem, USCG concluded that it was likely that illegal discharges had occurred. USCG determined that additional recordkeeping, in addition to solid waste management plans, was necessary to ensure compliance with the no-discharge rule for plastic and planned to provide boarding officers with better information.

USCG administers the regulatory program for the reception facility COA program, including periodic inspection of the port reception facilities to which those regulations apply (33 C.F.R. § 158). All port facilities and terminals under U.S. jurisdiction, which include "commercial fishing facilities, mineral- and petroleum-industry shore bases, and recreational boating facilities," must have garbage reception facilities (33 C.F.R. § 158.133; 33 C.F.R. § 158.400 et seq.). COA of garbage facilities is required for ports or terminals for oceangoing vessels of 400 gross tons or more carrying oily mixtures or noxious liquid substances, or fishing vessels that offload more than 500,000 pounds of commercial fish products a year (33 C.F.R. § 158.135). As of May 2008, approximately 2,295 facilities have valid COAs for MARPOL Annex V in the United States (U.S. Coast Guard, 2008).

MARPOL does not apply directly to warships or public vessels, although party states are to ensure that their public vessels operate reasonably and practicably within the Convention. While the U.S. Navy was not originally covered by APPS, MPPRCA amendments required that all ships, including those of the Navy, abide by the requirements of MARPOL Annex V by December 31, 1993.

Marine Protection, Research, and Sanctuaries Act of 1972

Title I of the Marine Protection, Research, and Sanctuaries Act (33 U.S.C. § 1401 et seq.), commonly referred to as the Ocean Dumping Act

(ODA), provides for the domestic implementation of the London Convention. EPA, the U.S. Army Corps of Engineers (USACE), NOAA, and USCG each have some authority under ODA. EPA regulates ocean dumping of all substances except dredged materials, which are under the authority of USACE, and is responsible for the designation of ocean dumping sites. NOAA and EPA are also authorized to carry out research. USCG is charged with maintaining surveillance and appropriate enforcement activities.

ODA requires a permit for any person seeking to transport material from the United States for the purpose of dumping in ocean waters. It also applies to persons seeking to dump from any vessel transporting material from outside the United States into a U.S. territorial sea or contiguous zone, and to U.S. flag vessels seeking to dump at any location. The ODA bans dumping of certain harmful wastes, such as radiological, chemical, medical, sewage sludge, and industrial wastes. Before getting a permit to dump other materials, the applicant must demonstrate that it "will not unreasonably degrade or endanger human health, welfare, or amenities or the marine environment, ecological systems or economic potentialities" (33 U.S.C. § 1412). The materials may be dumped only at designated sites, and USACE must use EPA-designated sites for disposal of dredged material to the maximum extent feasible. The vast majority of the materials dumped in ocean waters in the United States are sediments dredged from waterways to maintain navigation channels. Other materials, including fish waste, vessels, and human remains, are dumped in limited amounts and are subject to different requirements. ODA does not cover discharges into the territorial sea that are required to be permitted under the Federal Water Pollution Control Amendments of 1972 (33 U.S.C. § 1251 et seq.), commonly referred to as the Clean Water Act (CWA) (see later discussion), unless material is brought from outside the United States for dumping. EPA has promulgated applicable standards (40 C.F.R. §§ 220–229).

Unlike the London Convention, which specifically excludes the disposal at sea of wastes or other material incidental to the normal operations of a vessel (e.g., MARPOL Annex V garbage), ODA has a more limited exclusion from its definition of dumping for the discharge of effluent from the operation of propulsion or other motor-driven equipment or vessels. Thus, under domestic law, USCG has determined that, except for this limited exception, transportation of garbage as defined by APPS for the purpose of dumping at sea is covered by ODA (54 Fed. Reg. 18384, 18392 [April 26, 1989]).

As previously discussed, the London Convention was amended by the London Protocol to adopt a more stringent regime to protect the world's oceans from the impacts of ocean dumping. On September 4, 2007, the President submitted to the Senate his request for advice and consent for

U.S. ratification of the London Protocol. The legislative proposal to implement the London Protocol (i.e., amendments to ODA) was signed by the EPA Administrator on November 7, 2007, and transmitted to the U.S. Congress. Implementation of the London Protocol will not require significant changes to the U.S. ocean dumping program as it currently operates. Some changes to ODA would be needed, including incorporating the London Protocol's Annex I list of materials that may be considered for ocean dumping, except for sewage sludge, which is banned in the United States; extending coverage of ODA to the EEZ for dumping by non-U.S. vessels coming from outside the United States; prohibiting incineration at sea and export of waste for dumping or incineration at sea; prohibiting ocean dumping of low-level radioactive waste; explicitly covering subseabed storage and disposal; and requiring a permit for ocean dumping of fish wastes under certain circumstances. While the current Act does not prohibit incineration at sea or dumping of low-level radioactive wastes, it has long been U.S. practice not to authorize such activities.

Other Federal Marine Debris Laws and Programs

As described above, APPS, MPPRCA, and ODA are the primary domestic laws that implement MARPOL Annex V, the London Convention, the London Protocol, and the regulation and management of marine debris in the United States. However, there are many additional layers of laws that set out the international and domestic regulatory and management framework for the management of ocean- and land-based sources of marine debris. Several of these laws are described briefly.

Marine Debris Research, Prevention, and Reduction Act of 2006

The recently enacted Marine Debris Research, Prevention, and Reduction Act (MDRPRA) (33 U.S.C. § 1951 et seq.) established NOAA's Marine Debris Prevention and Removal Program; reconstituted the Interagency Marine Debris Coordinating Committee (IMDCC), which NOAA co-chairs with EPA; directed USCG to develop a strategy to improve implementation of MARPOL Annex V; and directed NOAA to establish a federal interagency marine debris data clearinghouse. The Act authorized up to $10 million for NOAA to implement the program, including mapping, identification, and impact assessments, removal and prevention activities, research and development of alternatives to gear that poses threats to the marine environment, and outreach activities; the Act also authorized $2 million for USCG activities. The MDRPRA, which was signed into law on December 22, 2006, authorized the NRC to prepare this report.

Clean Water Act of 1972

The Federal Water Pollution Control Amendments of 1972, more commonly known as the Clean Water Act (33 U.S.C. § 1251 et seq.), is important for the regulation of land-based sources of debris as a pollutant of marine waters and to the regulation of discharges from vessels. Under the CWA, discharging pollutants from point sources into the navigable waters and territorial seas of the United States is generally prohibited unless a permit has been issued consistent with applicable regulations issued by EPA. EPA has the authority to set higher standards for discharges into marine waters to prevent unreasonable degradation of ocean ecosystems (33 U.S.C. § 1251 et seq. § 403). The CWA defines "vessels and floating craft" as point sources. EPA has recently proposed a general permit to cover incidental discharges from normal operations of vessels to navigable waters of the United States (73 Fed. Reg. 34296 [June 17, 2008]). In discussing its proposal, EPA noted that the permit will not cover discharge of garbage that is regulated by USCG under MARPOL Annex V requirements.

Nonpoint source pollution and runoff during storms are the most significant sources of pollutants, including debris, that are washed into coastal and marine waters. Under Section 319 of the CWA (33 U.S.C. § 1251 et seq. § 319), states are required to develop management plans for controlling nonpoint source pollution. The plans must include identification of best management practices and measures and set annual milestones for implementation. EPA shares responsibility with NOAA and the states for implementing the coastal nonpoint source pollution control programs (see discussion of the Coastal Zone Management Act of 1972 [16 U.S.C. § 1451 et seq.]). Stormwater runoff, which is recognized as a major source of the marine debris along U.S. coasts and waterways, is regulated as a point source under the National Pollutant Discharge Elimination System. EPA's stormwater regulations require more than 5,000 municipalities, and many industrial facilities, to obtain permits (40 C.F.R. § 122.26). Most states are authorized to implement stormwater permit programs. The primary way for municipalities and industries to meet the stormwater regulations is to apply best management practices to prevent floatables and other pollutants from washing into storm sewers. EPA has also developed a combined sewer overflow control strategy, relevant mostly to older cities where sanitary and stormwater systems are combined, to minimize discharges and pollution of waterways and marine waters.

Some states have listed debris as a pollutant that is impairing their waters and are required to develop a total maximum daily load (TMDL) to identify measures to redress the impairment and the impacts of the debris (Environmental Protection Agency, 2002). TMDL is a calculation of the maximum amount of a pollutant that a waterbody can receive and still attain water quality standards, and an allocation of that amount to the pollutant's point and nonpoint sources (33 U.S.C. § 1251 et seq. § 303; 40 C.F.R. § 130.7). On their 1998 lists of impaired waters, California, New York, Alaska, Washington, and Connecticut identified a total of 62 water-bodies as water quality impaired because of debris, trash, floatables, and large woody debris. California has developed TMDLs for trash (California Regional Water Quality Control Board, 2001). In addition, under the leadership of its multiagency Ocean Protection Council, California has identified marine debris as a priority and has supported several initiatives to prevent and reduce marine debris (Box 3.2).

The Plant Protection Act and the Animal Health Protection Act

The Plant Protection Act and the Animal Health Protection Act (7 U.S.C. § 7701 et seq.; 7 U.S.C. § 8301 et seq.) are implemented and enforced by the U.S. Department of Agriculture's Animal and Plant Health Inspection Service (APHIS). Under the Federal Plant Pest Regulations (7 C.F.R. § 330.400–401) and the Animal and Animal Product Regulations (9 C.F.R. § 94.5), certain shipboard garbage that is aboard vessels entering U.S. ports from outside the United States (except Canada), as well as interstate movement of garbage between Hawaii and U.S. territories and possessions and other states, is regulated by APHIS to prevent the introduction or movement of plant pests and livestock or poultry diseases. Regulated garbage includes "all waste material that is derived in whole or in part from fruits, vegetables, meats, or other plant or animal (including poultry) material, and other refuse of any character whatsoever that has been associated with any such material" (7 C.F.R. § 330.400b; 9 C.F.R. § 94.5b). Among other requirements, regulated garbage is required to be kept in containers meeting specific requirements and may not be removed unless it is done under the direction of an APHIS inspector and sent to an approved facility. In order for a port or terminal to receive a COA for its garbage reception facilities, the port must certify that it is capable of receiving APHIS waste within 24 hours of receiving notices from an incoming vessel that it intends to discharge such waste (33 C.F.R. § 151.65 and 158.410). The civil and criminal penalties associated with violations of APHIS regulations and additional requirements associated with proper shoreside disposal of regulated garbage may result in an indirect incentive for vessels to dispose of plant- and animal-associated waste at sea in accordance with MARPOL Annex V prior to entering U.S. waters.

BOX 3.2
Marine Debris Initiatives: A California Case Study

The state of California has been a leader in the effort to combat marine debris. It has employed several methods to educate and involve the public. For example, the California Coastal Commission (2008) has a website to illustrate how trash becomes marine debris, how marine debris harms animals and humans, and how the public can help. The Commission also has an Adopt-A-Beach program, which works year round to collect marine debris washed up on California's beaches. The annual Coastal Cleanup Day, the highlight of the Adopt-A-Beach program, now receives more than 60,000 volunteers each year. Since the Cleanup Day program's introduction in 1985, over 800,000 volunteers have removed more than 12 million pounds of debris from California's coasts (California Coastal Commission, 2008). The California Ocean Protection Council, established in 2004 in accordance with the requirements of the California Ocean Protection Act, adopted a resolution to establish the following top marine debris priorities:

1. Reduce the sources of plastic marine debris.
2. Increase enforcement of anti-litter laws generally, and enforcement of laws to eliminate pollution by plastic resin pellets.
3. Seek innovative methods to reduce plastic waste.
4. Continue and expand watershed-based cleanups.
5. Increase the availability of trash, recycling and cigarette butt receptacles at public places, schools, and commercial establishments statewide.
6. Promote environmental education and outreach on the impacts of plastic debris and litter prevention.
7. Coordinate a marine debris steering committee.
8. Coordinate a regional effort.
9. Reduce single-use plastic packaging.
10. Remove derelict fishing gear.
11. Ban toxic plastic packaging.
12. Advance environmental education.
13. Prepare an education plan.
(California Ocean Protection Council, 2007)

The California Ocean Protection Council has also developed a draft strategy to implement this resolution (California Ocean Protection Council, 2008). This strategy pinpoints three policies that California should enact to eliminate marine debris: (1) create a "take-back" program, which would require manufacturers to reclaim and properly dispose of the plastic packaging from those products that would require such packaging; (2) establish a statewide ban on single-use plastic bags and polystyrene containers; and (3) enforce fees on other plastic packaging (California Ocean Protection Council, 2008). The state of California and several jurisdictions within the state have already enacted or are trying to enact legislation that speaks to these priorities (e.g., bans on some polystyrene food packaging, bills to reduce the use of single-use plastic and paper bags and to establish a reporting and removal program for derelict fishing gear).

Resource Conservation and Recovery Act of 1976

The Resource Conservation and Recovery Act of 1976 (RCRA) (42 U.S.C. § 6901 et seq.), administered by EPA in partnership with the states, sets out a comprehensive national framework for defining, managing, and regulating solid and hazardous waste. RCRA is important to proper management of shipborne waste when it reaches shore, including reuse and recycling. RCRA Subtitle D incorporates a solid waste management hierarchy that encourages waste minimization, reuse, and recycling efforts and provides for environmentally sound handling and disposal.

In 1987, the EPA Administrator was directed by Congress to undertake a study of the adverse impacts of plastics on the environment and methods to reduce or eliminate effects, particularly on fish and wildlife, as well as to conduct public outreach (42 U.S.C. § 6981). Recognizing the potential for plastic ring carriers to entangle fish and wildlife, in 1998 Congress also directed EPA to adopt regulations to limit this threat through redesign or requirements that the carriers be degradable in the marine environment (42 U.S.C. § 6914b and b-1).

The 1995 NRC report discussed at some length the need for an integrated vessel garbage management system that includes the existing land-based management systems and port and terminal operations (National Research Council, 1995a). Regrettably, there does not seem to be any significant progress toward this integration. For example, there still is no requirement for formal coordination between RCRA and the shipboard garbage management plans or port COAs. Responsibilities of ports and terminal operators for accepting garbage, providing onsite facilities for recycling, or coordinating with private contractors to ensure proper handling and disposal are still ill defined.

Coastal Zone Management Act of 1972

Under the Coastal Zone Management Act of 1972 (16 U.S.C. § 1451 et seq.), the federal government provides funding to the coastal states and territories to develop coastal zone management programs and enforceable policies consistent with national guidelines to enhance protection and conservation of the coastal environment, to support water-dependent businesses, and to guide development in the coastal zone. The states are required to conduct periodic self-assessments and to develop strategies to improve management in several areas, including the management of marine debris. States have developed numerous projects and programs to address marine debris; however, they vary widely and there are no specific requirements or measures of effectiveness. In 1990, Congress established the coastal nonpoint pollution control program under Section 6217 of the Coastal Zone Act Reauthorization Amendments of 1990

(16 U.S.C. § 1455b), which is jointly administered by NOAA and EPA. All states participating in the Coastal Zone Management Act must develop coastal nonpoint pollution management plans to implement measures largely consisting of best management practices under guidelines developed by EPA to control significant sources of polluted runoff in coastal waters (Environmental Protection Agency, 1993). One example relevant to marine debris is the establishment of Clean Marina Programs. The coastal nonpoint program was a driving force behind Clean Marina Program development in 12 states (Alabama, Connecticut, Louisiana, Ohio, Maine, Maryland, Massachusetts, Michigan, Mississippi, North Carolina, Texas, and Virginia) and helped to gain support for programs in many other states (see also Environmental Protection Agency, 2003).

Shore Protection Act of 1988

The Shore Protection Act of 1988 (33 U.S.C. § 2601 et seq.) was adopted to prevent deposition of municipal and commercial waste into U.S. coastal waters. It requires vessels that transport commercial or municipal waste in coastal waters to secure a permit from the Department of Transportation. Regulations promulgated by EPA establish minimum waste handling practices for vessels and waste handling facilities involved in the transport of municipal or commercial wastes in the coastal waters of the United States. These vessels and waste handling facilities are required to develop operation and maintenance manuals that identify procedures to prevent, report, and clean up discharges of waste in coastal waters. EPA has provided guidance on development of operation and maintenance manuals and encourages the use and documentation of existing industry practices that meet or exceed EPA's proposed minimum waste handling standards (40 C.F.R. § 237). The impetus for the Shore Protection Act was, in large part, due to the increased use of barges to transport waste and by the concomitant increase in floatables washing ashore in New York and New Jersey. As a response, an interstate and federal interagency effort was developed in 1989 for the New York/New Jersey Harbor complex. To date, approximately 350 million pounds of debris have been removed from the New York Harbor area. A summary and key findings from EPA's assessment of the floatables action plan is set out in Box 3.3.

Beaches Environmental Assessment and Coastal Health Act of 2000

Congress enacted the Beaches Environmental Assessment and Coastal Health Act of 2000 (33 U.S.C. § 1251 et seq.) to amend the CWA to reduce the risk of disease to users along coastal recreation waters. Under the Act, EPA is authorized to award grants to states, territories, tribes, and

---

BOX 3.3
Floatables Action Plan Assessment Report

The Floatables Action Plan was developed in 1989 to address floatable debris in the New York Bight, which includes the New York/New Jersey Harbor Complex and the shorelines of Long Island and New Jersey. The plan was developed jointly by an interagency workgroup that included representatives from the Environmental Protection Agency, the U.S. Army Corps of Engineers, the U.S. Coast Guard, the National Oceanic and Atmospheric Administration, the New Jersey Department of Environmental Protection, the New York State Department of Environmental Conservation, the New York City Department of Environmental Protection, the New York City Department of Sanitation, and the Interstate Sanitation Commission. The Floatables Action Plan has been carried out each year since to control wash-ups of floatable debris on area beaches. The plan consists of aerial surveillance via helicopter and fixed-winged plane, a communications network to report "slick" sightings and to coordinate cleanup response, and routine cleanups conducted by skimmer vessels in the harbor area. Since its inception, the plan has significantly reduced the amount of floating debris escaping the Harbor Complex and has expanded to include volunteer collection programs, boom and skim programs, combined sewer overflow collection programs, and beach cleanup programs.

2006 Floatable Observations

Twenty-one significant floatable slicks were observed in 2006. Newark Bay had the most slicks observed, nine, and the Kill Van Kull, with zero slicks observed, had the least. Six slicks were reported in the Lower New York Harbor, five slicks in the Upper New York Harbor, and one slick in the Arthur Kill.

Trends—Floatable Sightings in the New York/New Jersey Harbor Complex

A total of 513 significant slicks was observed over an 18-year period. The sightings of slicks have varied from year to year with the most number of slicks, 81, reported in 1990. The least number of slick sightings, 6 slicks, was reported in 1998. For unknown reasons, there was a significant increase in slick observations in 2004 followed by a decrease in 2005 and 2006. For the 13-year period, 53.6 percent of the observed slicks were in the moderate category, 26.8 percent were in the heavy category, and 19.6 percent were in the major category.

SOURCE: Environmental Protection Agency (2007a).

---

local governments to test and monitor coastal recreation waters adjacent to beaches or similar public access points. Grants can also be used to support programs to notify the public of the potential risk of exposure to disease-causing microorganisms in coastal recreation waters. EPA can also provide technical assistance to states and local governments to assess and monitor floatable materials. EPA issued a report entitled "Assessing and Monitoring Floatable Debris" that compiled and presented the most current information available addressing the assessment and monitoring of floatable debris (Environmental Protection Agency, 2002). This report provides examples of monitoring and assessment programs and mitigation actions designed to address the impact of floatable debris. It includes a specific discussion of plans and programs that seek to reduce floatable debris, including the New York/New Jersey Floatables Action Plan, EPA's combined sewer overflow program, and the Ocean Conservancy's International Coastal Cleanup Campaign.

Coral Reef Conservation Act of 2000

The Coral Reef Conservation Act of 2000 (16 U.S.C. § 6401 et seq.) established a coral reef conservation program at NOAA, required the development of a national coral reef action strategy, and authorized grants for coral reef conservation projects. At the time of passage of the Act, it was becoming increasingly clear that marine debris, particularly large amounts of DFG, were causing damage to fragile reef ecosystems. Under its Coral Reef Conservation Program, NOAA is authorized to provide "assistance to states in removing abandoned fishing gear, marine debris, and abandoned vessels from coral reefs to conserve living marine resources" (16 U.S.C. § 6401).

GAPS IN DOMESTIC REGULATION AND MANAGEMENT

More and more people are moving near the nation's coasts and the production of trash and floatable debris continues to increase. Without better control of the handling and disposal of trash and other wastes or reduction in the production and sale of items that are problematic as sources of pollution and waste, it is likely that the amount of such debris entering the nation's waterways will increase.

There currently is no national marine debris strategy and there are no clear lines of responsibility for addressing this multifaceted marine debris problem. While IMDCC holds out some promise for coordinating federal agency activities, there are still no clear lines of leadership, accountability, or commitment to funding program implementation over the long term.

Successful management of marine debris requires not only leadership and sustained funding, but also identification of the prevalence and potential sources of marine debris through monitoring and assessment programs, and also an understanding of the incentives, socioeconomic conditions, and activities that produce sources of marine debris. An important element of this effort is the effective integration of marine debris management and regulation into existing programs for coastal zone management, nonpoint pollution, and solid waste management. Reduction and abatement of the marine debris problems will require more extensive documentation and monitoring activities to assess the types, sources, and amounts of marine debris, combined with coordinated public education programs targeting key user groups and priority debris pollutants. This information is critical to developing effective solid waste management strategies focused on marine debris.

The effectiveness of laws and regulations to prevent ocean-based sources of marine debris will depend to a large extent on the ability and willingness of vessel operators to reduce the amount of potential waste coming aboard, to manage waste and garbage onboard their vessels, and to discharge their waste at ports and terminals that utilize good waste management practices. Involvement of academia, industry, and non-governmental organizations will be particularly important to understanding the full waste stream and identifying opportunities to reduce potential sources before they become garbage. Garbage management onboard ships and integration with the land-based solid waste management system pose many challenges that are not adequately addressed under the current laws and programs. This disconnect is reflected in the statement of task for this committee, which arbitrarily continues to separate analysis of ocean- and land-based sources. The gaps in domestic regulation and management can be addressed by improvement in the following four areas: (1) leadership and coordination; (2) integration of onshore and shipboard solid waste management systems; (3) enhanced interagency, industry, and public attention to waste minimization and source reduction; and (4) standards for enforcement and compliance. Finally, there is a need to develop programs and priorities for the mitigation and removal of debris that has been discharged into the marine environment.

Leadership and Coordination

In their recent reports, both the U.S. Commission on Ocean Policy and the Pew Oceans Commission highlighted the need for more effective interagency communication and coordination to improve the governance of U.S. oceans and coasts (Pew Oceans Commission, 2003; U.S. Commission on Ocean Policy, 2004). In response, the Executive Branch developed the U.S. Ocean Action Plan, which included a new interagency ocean governance structure to improve coordination among the myriad of federal agencies charged with overseeing the U.S. oceans and coasts, including revival of the IMDCC (Bush Administration, 2004).

Under several of the laws described above, including the most recent MDRPRA, Congress has charged federal agencies to coordinate efforts to address the problem of marine debris that has been historically spread across many agencies. Although the MDRPRA has established the NOAA Marine Debris Prevention and Removal Program and designates NOAA as the chairperson of the IMDCC, NOAA has designated EPA as co-chair of the IMDCC.

Congress has twice directed the formation of IMDCC. However, lack of consistent funding support and past efforts of IMDCC failed in effectively coordinating or galvanizing the interests and resources of key agencies to tackle the complex marine debris monitoring, assessment, and mitigation challenges. The 1995 NRC report made the call for national leadership a keystone of their recommendations, going so far as to call for a permanent national marine debris commission (National Research Council, 1995a). It is not surprising that marine debris has not consistently received high priority given the complex framework of laws and agency responsibilities, and agencies' more prominent mission mandates such as homeland security, toxic pollution and hazardous waste reduction, and atmospheric research and fisheries management.

This failure also reflects the historical and legal boundaries of authority imposed on these agencies by statute and their different and sometimes rigid institutional cultures. These differences are exacerbated when budgets are tight and collaboration takes a back seat to core mission responsibilities. USCG manages ships and U.S. ports, including port waste reception facilities, but does not have principal responsibility for overall solid waste management. EPA manages a wide range of pollution sources and pollutants, but has limited authority over maritime activities. USDA is responsible for inspection of certain plant and animal wastes but is not involved in regulating other shipborne garbage. NOAA manages living marine resources, including their habitats, and has been given additional marine debris program responsibilities under the MDRPRA—including prevention, assessment, research, and outreach—but has limited authority over ships, polluters, and shoreside coastal activities in general. The U.S. Fish and Wildlife Service manages some marine mammals and other coastal wildlife and birds and has jurisdiction over U.S. National Wildlife Refuges, but it has no direct role in controlling marine debris on land or at sea. The U.S. National Park Service (NPS) is responsible for maintaining the integrity of National Parks and Seashores; but, like the U.S. Fish and Wildlife Service, NPS has no direct role in controlling marine debris on land or at sea. In addition, USACE has responsibility over dredging and filling in the coastal zone but lacks authority over ocean- or land-based sources of marine debris.

Agencies have assessed their own roles within their own authority and in some cases have had limited success (Donohue, 2003; Sheavly, 2007). As previously described, progress being made by agency and interagency actions appears to be local, ephemeral, and in some cases upstream from direct deposition or impacts (e.g., vessel placards, public outreach, port COAs). A review of some past efforts by NPS, NOAA's National Marine

Fisheries Service, and EPA is provided by Ribic et al. (1997). Unfortunately, progress on various subelements of the marine debris problem does not constitute progress on the overall problem. As suggested by NRC (1995a), this vexing situation requires either more effective interagency cooperation or the creation or designation of a senior agency to fill this role. Until there is clear direction of a lead agency responsible for addressing marine debris as a priority issue or for addressing gaps in the current regime and assuming responsibility for interagency coordination, it will be very difficult to manage the marine debris problem comprehensively, effectively, and for long enough to be successful. Consideration needs to be given to ensuring long-term support of NOAA's new Marine Debris Prevention and Removal Program and to clarifying the roles and responsibilities of NOAA and other agencies in IMDCC.

Finding: Although the U.S. Congress has charged federal agencies with addressing the marine debris problem and has called for interagency coordination, leadership and governance remain diffuse and ineffective.

Recommendation: IMDCC or Congress should clearly designate a lead agency to expand cooperative marine debris programs, including but not limited to land-based marine debris, DFG, shipborne waste, and abandoned vessels. IMDCC should develop a national strategy and national standards and priorities for dealing with all elements of marine debris. The strategic plan should include a clear identification of lead agencies, an implementation schedule, and performance benchmarks.

Finding: There is no formal or functional coordination between RCRA (42 U.S.C. § 6901 et seq.), which regulates U.S. waste management and disposal, and the shipboard solid waste management plans or port and terminal waste management and COAs.

Recommendation: Specific performance standards should be developed by USCG in collaboration with EPA for COAs; approval of port COAs should be conditioned on formal coordination between ports and solid waste management systems based on the RCRA waste management hierarchy and best management practices and guidance developed by EPA. Performance standards and COA and port discharge requirements should be based on an understanding of the capacity and capabilities of vessel types and waste streams, not just a hypothetical capability to handle wastes. The private sector and nongovernmental organizations should be included as partners in these efforts.

Integrated Solid Waste Management System

As previously described, both nationally and internationally, there continue to be obstacles and disincentives to proper landside disposal of waste generated at sea (see "Port Reception Facilities" discussion earlier in this chapter). While there have been repeated calls for an integrated ship-to-shore solid waste management system (e.g., National Research Council, 1995a), this remains a significant challenge. Port facility review is further complicated by the varied (and in many cases very limited) roles that terminal operators and port authorities play at the vessel–port interface of the solid waste flow. In some cases, the ports are no more than silent partners where ship operators contract directly with waste management firms. Port authorities, who are not in the garbage business, are reluctant to take on a direct management role or take responsibility for ship-generated solid waste, and local solid waste management program operators have little incentive to incorporate ship-generated solid waste into their management programs. EPA needs to work actively with states, ports, terminal operators, and the private sector to increase collaboration between RCRA and CWA program offices and to identify approaches and support state efforts to incorporate solid waste streams from ships into the local and state solid waste management plans.

While much of the focus of past marine debris efforts has been on ships and vessels under the jurisdiction of MARPOL Annex V and APPS, there is also a need to expand programs targeted at debris from other and smaller vessel types, as well as COAs or similar certification for smaller ports. USCG, EPA, and NOAA, as well as industry and nongovernmental organizations, have education and outreach programs directed toward recreational boaters and other vessels that could be expanded.

Finding: Despite past recommendations and legislative mandates for collaboration, there continues to be a legal disconnect and jurisdictional discontinuity between solid waste management mandates afloat and ashore.

Recommendation: EPA should work with state and local solid waste management programs and port and terminal operators to support a seamless connection and accountability for transfer of ship-generated garbage into the terrestrial waste management system.

Waste Minimization and Source Reduction

There has been substantial progress in ship-generated solid waste management practices since the adoption of MARPOL Annex V and its implementing legislation. More effective implementation of vessel-based waste management can be improved by

- adoption of prevention and mitigation programs to reduce the sources of marine debris that are used in ordinary ship operations;
- changes in waste-handling practices and technology onboard vessels to include waste reduction and recycling and to incorporate zero-discharge goals where feasible and practicable; and
- expanded efforts to ensure the adequacy in fact, the implementation of cost-effective enhancement of port reception facilities, and the integration of shipboard and onshore solid waste management plans and systems.

Industry and nongovernmental organizations have also taken an interest in the development of environmental management systems that can enhance waste management. For example, the International Organization for Standardization (ISO) has developed standards and guidelines for environmental management systems (ISO 14000). Application of ISO 14000 or similar environmental management systems to port operations, ship operations, and the ship–port interface could minimize waste and reduce marine debris (e.g., Urban Harbors Institute, 2000; Environmental Protection Agency, 2004).

Segments of the cruise and ocean shipping sectors provide good examples of effective shipboard solid waste management programs. For example, members of the Cruise Lines International Association, Inc., have adopted mandatory environmental standards with a goal of zero at-sea discharge of solid wastes and overall waste minimization procedures (Cruise Lines International Association, Inc., 2006; Environmental Protection Agency, 2007a). The Cruise Lines International Association's programs also include stringent monitoring and auditing practices and procedures. Similarly, the Matson Navigation Company has implemented a "zero solid waste discharge" program for its domestic containership route. Under this program, Matson has limited waste disposed at sea to food scraps; all other solid waste materials are retained for recycling or disposal at shoreside facilities. Zero-discharge initiatives made possible by the implementation of aggressive solid waste minimization programs have been implemented in shoreside operations for a number of years and, although some of the approaches and technologies they have developed may not be applicable to vessel operations, other approaches may be transferable.

Finding: Zero discharge, source reduction, and waste minimization practices have been implemented in industrial settings ashore for a number of years. Some vessels have successfully adopted zero or minimal discharge practices based on these successful shoreside models.

Recommendation: USCG, in coordination with EPA, should promulgate best management practices that reflect the maximum practicable extent to which ships can operate without the need to dispose of garbage at sea. Development of these best management practices should be based on successful zero discharge, source reduction, and waste minimization practices, coupled with an understanding of the technical and financial abilities of different vessel types to retain different forms of waste. IMDCC should support the adoption of voluntary zero waste discharge standards and implementation of these best management practices to achieve that goal.

Source reduction efforts will require public–private partnerships and the active involvement of manufacturers, industry groups, ports, and solid waste management agencies to be successful. Industry efforts to reduce overall amounts of packaging and to develop more environmentally friendly materials, including reduction of plastic trash and increased recycling of plastics, can potentially contribute to significant reductions in marine debris. One example is the American Chemistry Council marine litter campaign, which brings leaders from government, academia, industry, and nongovernmental organizations together to develop solutions. In 2007, the American Chemistry Council held a workshop, in conjunction with the Scripps Institution of Oceanography, to discuss current research, sources of marine debris—both land- and ocean-based—and possible solutions. These efforts were started partly in response to rising public concern in California and other states about plastics pollution, and local and state efforts to limit and regulate plastic packaging. It will be increasingly important for all agencies, academic institutions, industries, nongovernmental organizations, and other stakeholders to support public and private partnerships to effectively tackle specific marine debris problems.

Finding: There is a need to focus additional attention on potential waste before and after it reaches the ship. Shipboard-focused programs are unlikely to be fully successful without additional efforts to encourage source reduction on the front end and additional efforts to ensure reception facilities consistent with projected needs on the receiving end.

Recommendation: EPA should take the lead in coordinating with IMDCC to work with academia, industry, and nongovernmental organizations to develop industry standards and guidelines for source reduction, reuse, and recycling for solid wastes that are utilized and generated during normal ship operations.

Enforcement and Compliance

In Chapter 2, the committee noted the limited amount of quantitative monitoring and research data available; however, these data suggest that measures taken thus far have not been successful in abating the problems. Similarly, there is difficulty in assessing the effectiveness of the regulatory framework at both national and international levels; information that is available indicates that the effectiveness could be improved. The committee received presentations from a number of experts which indicated that existing metrics of effectiveness are limited and, in most cases, nonexistent from a global perspective.

Recordkeeping (e.g., number of vessels discharging garbage at ports, amounts of garbage discharged, number of reports of inadequacies, number of violations) could all be useful indicators of effectiveness. Yet, there is no comprehensive collection of this type of data domestically or internationally. Garbage management plans and logbooks provide only a vague idea of compliance and do not apply to vessels smaller than 40 feet. The number of port COAs or reports of inadequacies may be an indicator of the ability (or inability) of vessels to discharge their waste shoreside. For example, EPA's "Draft Cruise Ship Discharge Assessment Report" examined, among other issues, the USCG port reception facility COA program (Environmental Protection Agency, 2007b). The report noted that USCG conducted over 14,000 facility inspections in 2006, up from approximately 3,500 inspections conducted during calendar year 2000. These included inspections of MARPOL Annex V port reception facilities for compliance and adequacy. USCG issued or responded to and investigated 2,986 complaints of reception facility deficiencies in 2006, up from 2,587 in calendar year 2000. The report also noted that, from 2002 to 2006, USCG has documented a 26 percent reduction in the number of pollution incidents reported at facilities. However, these and similar analyses are done a posteriori, without accounting for confounding issues that may contribute to changes in compliance and adequacy. They provide limited insight into the effectiveness of MARPOL Annex V.

Enforcement data are also interesting but inadequate for assessing effectiveness because enforcement actions may be indicators of enforcement effort, or even happenstance, rather than accurate indicators of noncompliance rates. Therefore, it was difficult for the committee to assess the effectiveness of international and national measures to prevent and reduce marine debris based solely on regulatory information. Nevertheless, the continued presence of ship-generated waste in the marine environment clearly indicates that challenges to MARPOL Annex V implementation have not been overcome. A meaningful understanding of the efficacy of these regulations is required by decision makers if improvements are to be achieved through enhanced and refined regulatory language, increased compliance, outreach, and other means. Within the United States, the demands on USCG for oversight and enforcement will increase as ship-borne commerce expands. There is concern that USCG does not have sufficient resources and trained personnel to ensure a fully effective marine solid waste management regime. The need for trained resources will be more important if future efforts expand beyond review of operational compliance with garbage manifest regulations to consideration of qualitative standards for proper solid waste management, both on the vessel and at the port.

The second edition of the *Guidelines for the Implementation of Annex V of MARPOL* explicitly addresses compliance issues:

> Recognizing that direct enforcement of Annex V regulations, particularly at sea, is difficult to accomplish, governments are encouraged to consider not only restrictive and punitive measures but also the removal of any disincentives, creation of positive incentives, and the development of voluntary measures within the regulated community when developing programs and domestic legislation to ensure compliance of Annex V (International Maritime Organization, 2006b).

These guidelines on enforcement, compliance incentive systems, and voluntary measures provide opportunities for national and international data collection and analysis. The NRC (1995a) outlines the enormity of developing a national data system and devotes an entire section in the report on recordkeeping as a measure of MARPOL Annex V implementation. The findings and recommendations from Chapter 8, Measuring Progress in Implementation of MARPOL Annex V, remain valid and largely unexecuted. To turn IMO's guidelines into mandatory practices would be a significant step forward.

Finding: Forensic analysis of enforcement and compliance information is a necessary tool for evaluating the effectiveness of implementation of MARPOL Annex V; however, there is no comprehensive system in place for collecting and analyzing information for this purpose at either domestic or international levels.

Recommendation: USCG, in coordination with IMDCC, should develop a program to analyze the effectiveness of domestic regulations to reduce marine debris. Where feasible, it should utilize recordkeeping, enforcement, and other data that are already being collected and should investigate additional metrics that may be useful in measuring effectiveness. The U.S. delegation should recommend that IMO, in its ongoing review of MARPOL Annex V, incorporate this program into a global analysis of the effectiveness of MARPOL Annex V.

Debris Mitigation and Removal

It is readily apparent that there is no national strategy for mitigation and removal of marine debris. A national strategic plan would identify the aspects of marine debris that are most troublesome. For example, to what extent is society concerned with and impacted by visual disamenities associated with littered shores, health and safety issues related to hazardous wastes or pollution caused by debris, or the various ecosystem impacts or impacts on species at risk? A meaningful strategic plan cannot be developed without first prioritizing concerns, as well as identifying opportunities for taking action. Next, a comprehensive inventory of the current spatial distribution of littoral, benthic, and pelagic debris and its composition is needed, as well as knowledge of debris sources. Then, estimates of the costs of prevention, mitigation, and removal are needed. Armed with these types of information, society would be well positioned to steer limited resources to the most cost-effective projects—projects that address priorities at the least cost. There are numerous examples of this approach in business and in government. The EPA Superfund program is an example of how to prioritize sites and to methodically remediate them. A risk assessment approach (National Research Council, 1983, 1993, 1995b, 1996b, c, 2002, 2004; Pratt et al., 1995) could serve as an alternative template for prioritizing mitigation and remediation projects. In the case of marine debris, it is likely that society has a plurality of objectives that is not neatly subsumed into a strict hierarchy; in this case, there are a variety of multicriteria decision-analysis approaches that could be brought to bear to prioritize mitigation and remediation projects (e.g., Keeney and Raiffa, 1976; Saaty, 1990; National Research Council, 2004). There may be some sites that are so expensive to remediate (e.g., the deep ocean) that the funding necessary to launch a remediation project can never be advanced. There may be some sources of debris generation that are so ubiquitous, diffuse, or otherwise difficult to control in the marine environment that it is preferable to examine ways to limit their production or sale or alter production processes. While these "upstream" solutions are beyond the scope of the report, for the most problematic and pervasive types of garbage, they merit greater attention in the future by EPA and other federal and state agencies working with the private sector.

Selecting priorities for prevention, mitigation, or remediation is only a first step. It is also necessary to devise an incentive structure to support the achievement of objectives. There are four fundamental approaches to influencing behavior: (1) moral suasion or social pressure; (2) standards or mandated technological solutions; (3) fines, taxes, or subsidies; and (4) definition and allocation of entitlements and obligations within a market structure. The design of an incentive structure to support the prevention of continued deposition of debris into the marine environment and the mitigation or removal of marine debris that is present may use any combination of these incentive systems. Public awareness campaigns can shame people into being less inclined to drop their trash on the beach or overboard. Clean Marina and Clean Harbor programs are examples of moral suasion campaigns. Fines or subsidies can lead people to desist from undesired behaviors or to engage in desired behaviors. However, to be effective there has to be a reasonably high probability of the adverse behavior being detected so that the expected value of the fine is meaningfully large. Taxes change effective prices and lead people to change their behavior to minimize their tax burden and can be an effective way of influencing choices of production and consumption technologies and behaviors. For example, a tax on synthetic fishing gear would encourage a more conservative use of synthetic gear and, if high enough, might lead to reconsideration of biodegradable fibers. Defined standards or technologies have been used by EPA for emissions control. In general, defining required performance standards and leaving people free to determine how to achieve those standards results in higher compliance and lower costs of compliance. Definition and allocation of entitlements and obligations within a market structure work well for activities that are easily observed by other participants, even if they are not easily observed by enforcement agencies. However, the deposition of marine debris is not easily observed. If it were, beaches would not fill with litter and vessels would not complete transoceanic voyages without needing to dispose of at least some plastic debris in their port of call. When there are multiple objectives to attain, as is likely the case for a national marine debris strategic plan, it is likely that a combination of incentive structures will be needed.

Finding: Current marine debris mitigation efforts are episodic and crisis driven. There is a need for a reliable, dedicated funding stream to support marine debris mitigation efforts and a national strategy and framework for identifying priorities for removal of marine debris.

Recommendation: IMDCC should work with nongovernmental organizations and the private sector to identify and establish a national strategic plan for addressing the marine debris problem and to identify funding mechanisms and reliable funding streams to support marine debris mitigation activities.

CONCLUSION

The following findings and recommendations express overarching concepts discussed in the previous findings and recommendations in Chapter 3.

Overarching Finding: Despite measures to prevent and reduce marine debris, evidence shows that the problem continues and will likely get worse. This indicates that current measures for preventing and reducing marine debris are inadequate. At both the international and the domestic levels, marine debris responsibilities and resources have been spread across organizations and management regimes, slowing progress on the problem. Improvements will require changes to the regulatory regime as well as nonregulatory incentives. At both the international and the national levels, there needs to be better leadership, coordination, and integration of mandates and resources.

Overarching Recommendation: The United States and the international maritime community should adopt a goal of zero discharge of waste into the marine environment. The United States should take the lead in the international arena in this effort and in coordinating regional management of marine debris with other coastal states. IMDCC should develop a strategic plan for domestic marine debris management. Performance measures should be developed by the United States and the international maritime community that allow for assessment of the effectiveness of current and future marine debris prevention and reduction measures.

Overarching Finding: The lack of understanding of vessel waste streams and the inadequacy of port reception facilities to accept and properly manage vessel waste is a serious impediment to the prevention and reduction of marine debris, including DFG. Ships continue to face shoreside disposal challenges at some berths in countries that have formally communicated the availability of adequate reception facilities.

Overarching Recommendation: To achieve the goal of zero discharge, ships need to be able to discharge their waste at ports and should have incentives (or at least they should not face disincentives) to do so. Domestically, USCG should

- establish minimum qualitative and quantitative standards for port adequacy,
- provide technical assistance for ports to achieve standards,
- encourage ports to provide incentives to vessel operators for discharging their waste ashore, and
- ensure that there are adequate reception facilities and alternative disposal options (see Appendix E) for waste fishing gear.

Internationally, the U.S. delegation to IMO should exert its leadership in the ongoing MARPOL Annex V review process to ensure that similar amendments are incorporated into Annex V.

4

Derelict Fishing Gear and Fish Aggregating Devices

While all maritime sectors, from recreational boats to large commercial shipping vessels, contribute to the ocean-based marine debris problem, there has been growing concern about the contribution of fishing vessels to this problem. Endangered monk seals entangled in derelict nets in the Northwestern Hawaiian Islands (NWHI) and tons of fishing gear being hauled away from remote Alaskan shorelines are vivid evidence of the serious nature of this problem. Commercial fishing vessels generate a significant portion of the U.S. maritime waste stream, including waste fishing gear (Cantin et al., 1990; National Research Council, 1995a). Both derelict fishing gear (DFG) and fish aggregating devices (FADs) were specifically referenced in the Marine Debris Research, Prevention, and Reduction Act (33 U.S.C. § 1951 et seq.) as subjects for further review by this committee. While they can be marine debris, there are legal and practical considerations that differentiate them from other debris types. And in some coastal areas, a very large proportion of marine debris is often related to fishing (e.g., northern Australia [Kiessling, 2003], NWHI [Donohue et al., 2001], Aleutian Islands [Merrell, 1980, 1984, 1985]). Therefore, the committee has chosen to devote a separate chapter to exploring these types of debris. This chapter begins by examining DFG and follows with a discussion of FADs, which become DFG once they are abandoned.

DERELICT FISHING GEAR

Arguably the single most important advancement in fisheries technology is the replacement of natural, easily degraded fiber ropes and twines with cheap, durable, and lightweight synthetic ropes and twines (Kristjonsson, 1959). Historically, hemp, cotton, jute, sisal, manila, silk, and linen were the primary natural fibers used to make fishing gear (Uchida, 1985; Brainard et al., 2000). They were treated with a wide variety of dyes, tars, and preservatives to retard their rate of degradation in the marine environment. Nevertheless, their failure, replacement, and repair rates were very high. These strength and durability limitations were major factors that limited catch sizes in many fisheries. Advances in polymer chemistry and production technology in the post–World War II period led to the manufacture of polyethylene, polypropylene, polyamide (nylon), and other synthetic fibers which have all but replaced the natural fibers used in fishing gear. Worldwide, these advances greatly contributed to the vast growth in fish and shellfish harvesting capacity and also set the stage for resource management challenges that are yet to be fully and effectively addressed by governments and industry. While achieving sustainable fisheries is still the primary challenge of management authorities, another result of this technological revolution that has largely been overlooked is the effect of the loss or discard of these persistent materials into the marine ecosystem. The same properties that make these new materials effective as fishing gear also make them particularly problematic as marine debris. Unlike their natural predecessors, the new materials can last for years or decades in the marine environment. They are largely impervious to biodegradation; they are resistant to chemicals, light, and abrasion; and because many of these synthetic fibers are buoyant, they can be transported long distances by ocean currents.

With the entry into force of the International Convention for the Prevention of Pollution from Ships, 1973, as modified by the Protocol of 1978 (MARPOL) Annex V and its implementation via domestic laws in the late 1980s, the at-sea discharge of plastics and other synthetic polymers, including fishing gear, was prohibited. This change from long-standing ship disposal practices, coupled with concurrent rising public awareness of synthetic materials–based marine debris (e.g., Manheim, 1986; Adler, 1987; O'Hara et al., 1988; Toufexis, 1988), increased focus on the problems associated with fishing gear lost or discarded into the marine environment. DFG is of particular concern because the use of synthetic materials has made fishing gear more durable and because it can continue to entrap, entangle, and retain marine organisms after it has been lost or discarded.

The committee defines fishing gear as any device or equipment or parts thereof, except vessels, used in the catching, attracting, gathering, holding, and harvesting of marine or aquatic species. DFG is fishing gear in the marine or littoral environment that has been abandoned or is otherwise no longer under the control (in the context of the legitimate operations of the specific fishery) of its legal operator. This definition does not address the many circumstances that may result in the loss of control of fishing gear, but it recognizes that what constitutes "control" of fishing gear varies among specific fisheries and fishery management systems. The term "derelict" refers to the intentional or unintentional abandonment of the gear. In either case, the operator acknowledges that he must abandon (relinquish control of) his gear; hence, the use of the term "derelict" fishing gear is appropriate. Fishing gear is unlike most other discharges or disposals considered in MARPOL Annex V in that it is intentionally deployed into the marine environment with the intention of retrieval. Commercial fishing gear is capital equipment used in the pursuit of value associated with the trade in fisheries products. In deciding to deploy their gear, fishermen engage in an implicit balancing of the expected value of their catch and the risk of damaging or losing their gear (Pooley, 2000). The quality of these judgments varies with experience; environmental conditions (e.g., weather, currents, tides, sea state, presence of sea ice, the makeup of the seafloor); the condition of the gear, equipment, and vessel; as well as a suite of economic pressures and regulatory factors. The fact is that fishing, legal or otherwise, entails risking the loss of some fishing gear. The challenge for fishermen, fishery engineers, fishery managers, and lawmakers is to find ways to incorporate the minimization of gear loss and its ultimate environmental hazards and the maximization of lost gear recovery into fishing operations, research programs, management and enforcement actions, and public policy directions.

Sources, Fates, Abundance, and Impacts

Prevention and reduction of DFG and its impacts requires an understanding of the sources, abundance, and impacts of this gear. While this information is also discussed in Chapter 2, some of what is known and some of the challenges in understanding DFG are highlighted here. As is true for other types of marine debris, there is little information available in the form of quantitative assessments of the sources and amounts of derelict gear generated by specific fisheries, or for linking those losses to impacts.

Prevention of DFG begins at the source, but identifying the source may be difficult because ocean currents can transport DFG a long distance from the site of loss or discard and involve substantial time lags (Donohue et al., 2001; Boland and Donohue, 2003; Kubota, 1994; Donohue, 2005; Kubota et al., 2005; Pichel et al., 2007). As such, DFG encountered within the U.S. Exclusive Economic Zone (EEZ) or on U.S. shorelines may be derived from current and past activities of domestic and foreign fishing fleets operating within or beyond the EEZ. There is evidence that fishing gear manufactured in Asia, particularly South Korea, Japan, and Taiwan, represents a significant component of DFG recovered in Alaska, Hawaii, and northern Australia (Kiessling, 2003; White et al., 2004; Timmers et al., 2005; Carpentaria Ghost Nets Programme, 2008; Bob King, personal communication; Michael Stone, personal communication). Much of the DFG documented in these locations is composed of materials commonly manufactured in Asia (e.g., *twisted* polyethylene twine); these materials are (reportedly) rarely used by manufacturers of fishing gear currently used in the United States, who instead use netting of domestic origin and from the European Union (e.g., *braided* polyethylene twine produced in Iceland and Portugal) (Bob King, personal communication; Michael Stone, personal communication). Complicating this situation is the existence of "legacy" gear; some derelict gear recovered in Alaska and Hawaii is very old, suggesting it may represent a relic of foreign fishing in what are now U.S. waters (Bob King, personal communication; Michael Stone, personal communication; and see Merrell, 1980). Prevention and reduction of DFG will have to take into account the transport of these materials across boundaries over long periods of time.

> **Finding:** Because DFG persists and can be transported long distances, parties that generate DFG may not be the ones that bear the effects of it. Increased awareness and participation by responsible parties is necessary to effectively address the DFG problem.

> **Recommendation:** All parties responsible for the generation of DFG should be involved in prevention and cleanup. Measures to prevent and reduce DFG will require international coordination and cooperation. The National Oceanic and Atmospheric Administration (NOAA), the U.S. Department of State, international fisheries management organizations, and other relevant organizations should
>
> - engage in technology transfer and capacity building with nations from which DFG components originate to improve implementation of MARPOL Annex V in fisheries;
> - encourage best practices to reduce gear loss, support recycling of used fishing gear, and promote retrieval of snagged or lost gear; and
> - facilitate the participation of representatives from nations from which DFG components originate in DFG survey and removal efforts.

While the origin of some legacy gear is uncertain, sources are clearly identifiable in other cases; for example, the Northwest Straits Commission estimates that there are nearly 3,900 gillnets remaining in Puget Sound from domestic salmon fisheries from the 1970s and 1980s (Natural Resources Consultants, Inc., 2007). Ongoing domestic fisheries also contribute to derelict crab and lobster pots in the Atlantic and Pacific Oceans, the Gulf of Mexico, and Alaska (e.g., Hess et al., 1999; Guillory et al., 2001; National Oceanic and Atmospheric Administration, 2008b; Thomas Matthews, personal communication; Steven Vanderkooy, personal communication). In northeastern Atlantic fisheries, the amount of lost and discarded nets is unknown, but anecdotal evidence suggests that, in some fisheries, 30 km of net are lost or discarded during a typical 45-day trip, which translates into 1,254 km of lost or discarded netting per year (Hareide et al., 2005).

It is also important to note that gear types and materials are constantly evolving; in considering measures to prevent and reduce DFG and its impacts, it is crucial to consider new fishing technologies and how these technologies may affect fishing behaviors. Most active fisheries are continually searching for materials more suited to fishing needs. For example, trawlers are currently exploring the use of aramid fiber–based netting, which is extremely strong, lightweight, and abrasion resistant. These new fibers sink and are less likely to tear apart when snagged or heavily loaded during fishing. This may be a positive development with respect to gear loss, but the degree to which improved materials leads to higher levels of (gear loss) risk taking by fishermen is not known. Typically, improvements in any fishing technology and techniques are aimed at catching more fish at less cost and those that may coincidentally reduce the probability of gear loss may also affect fishing behavior so as to cancel those benefits (Coe, 1990).

DFG has been recognized as a particularly hazardous form of marine debris since the earliest reports on effects of persistent waste materials in the environment. Records of the entanglement of threatened and endangered species, such as sea turtles, fur seals, Hawaiian monk seals, some large whales, and many seabird species date back to the 1970s (Gochfeld, 1973; Bourne, 1976, 1977; Balazs, 1978). Ghost fishing has been confirmed to occur with many static types of fishing gear (e.g., gillnets, traps, baited hooks), with potentially significant impacts on commercial stocks in some fisheries (Breen, 1987; Stevens et al., 2000; Sancho et al., 2003; Matsuoka et al., 2005; Brown and Macfayden, 2007). Other forms of DFG also have the potential to entangle marine organisms, disable vessels, cause physical damage to habitat, and contribute to the marine debris problem.

The word "fishing" encompasses a broad range of activities pursued with a variety of equipment; therefore, solutions to prevent and reduce

DFG must be tailored to the different types of gear, their impacts, and the primary causes of loss. This section describes the primary types of fishing gear that can become derelict and includes a brief summary of the impacts and causes for loss by gear type.[1]

Trawl Nets

Trawl nets (trawls) are expensive funnel-shaped nets towed by one, or sometimes two, vessels through aggregations of fish. Trawls can be designed to fish anywhere in the water column, from contact with the seafloor to the middle or upper portions of the water column. Fish that are herded into the mouth of the net are eventually concentrated in the end of the funnel, or cod end, and winched aboard the vessel. Trawl designs vary depending on the target species, vessel size, and regulatory limitations. Originally, trawl webbing was made from hemp treated with various preservatives, but these were replaced with nylon and then polypropylene and polyethylene in the 1950s and 1960s (Uchida, 1985) and more recently by next-generation polyethylene fibers such as Spectra® and Dyneema® and aramid fibers such as Kevlar® (Michael Stone, personal communication). Most of these materials float, which accounts for the fact that derelict trawl webbing and cod ends from trawl nets are found worldwide and notably in concentrations on shores up to thousands of miles from their putative origins (Merrell, 1980; Henderson et al., 1987; Donohue et al., 2001; Kiessling, 2003; White et al., 2004). Even though sections of trawl webbing may be buoyant, steel cables, doors, beams, and other materials used to maintain the vertical and horizontal profiles of the trawl may be weighted and the trawl as a whole is negatively buoyant. The increased strength of synthetic webbing reduces hydrodynamic drag and enables vessels to pull larger nets at higher speeds and greater depths.

Historically, the loss of trawl gear was attributed mainly to snagging and tears while fishing near or on the bottom. In U.S. domestic trawl fisheries, fishermen have stated that this type of net loss is less common (Bob King, personal communication; Michael Stone, personal communication). Part or all of these nets can be lost in a snagging incident, and the repair process may generate waste webbing and lines that may be discarded or lost overboard. Trawl webbing has been identified in the entanglement of seals and sea lions (see Appendix C, Table V). It is widely distributed on coasts from tropical to Arctic and Antarctic regions of the world (Merrell, 1985; Uchida, 1985; Ryan, 1990; Ribic et al., 1992; Boland and Donohue, 2003). Trawl webbing has also been identified as particularly destructive

---

[1]Additional information on fishing gear technology can be found through the Food and Agriculture Organization of the United Nations (2008a).

to fragile coral reef systems in the NWHI and similar habitats in Australia (Donohue et al., 2001; Kiessling, 2003; White et al., 2004).

Gillnets

Gillnets are vertical walls of mesh, sized such that target species in the desired size range are effectively caught about their girth after the head and gill covers (fish) have passed through the mesh. Gillnets are cheap to manufacture and are used worldwide in fisheries that vary from artisanal to industrial in scale. They are efficient size selectors for the target species, but they are also effective at ensnaring nontarget species, especially those that have heads small enough to pass through the mesh or that have prominent spines and angular carapaces (Carr et al., 1985; High, 1985; Breen, 1990). To maintain their shape, gillnets usually include a buoyant top (cork) line and a weighted bottom (lead) line, and they can be suspended at or near the surface, in midwater, or anchored (set) to the bottom. Early gillnets were made from natural fibers such as cotton. Current gillnets are woven from mono- and multifilament nylon, Dacron® twine, and Spectra®. Gillnets can vary in size from as small as a few square meters up to systems of nets as long as 60 km by 20 m deep, with mesh sizes varying from as small as 2 cm to as large as 50 cm. Primarily because of concerns over bycatch and ghost fishing, management interest in controlling these fisheries and their gear loss rates has been high (e.g., coastal state bans, high seas bans). Despite the United Nations ban on large-scale high seas drift gillnets and similar multilateral treaties, drift gillnets continue to be used (Brainard et al., 2000; Food and Agriculture Organization of the United Nations, 2008b).

Gillnets in coastal waters are most often lost due to snagging or when attempts to retrieve them cause tears. Sections of gillnet are often lost or cut away when they become entangled as vessels jockey for position in derby fisheries such as the Bristol Bay sockeye fishery. Floating or drift gillnets are lost when marker buoys are lost in foul weather or are entangled or carried away by vessels that transit through them, or when the weight of their catch causes them to sink. In addition, an unintended consequence of prohibitions on the use of high seas drift gillnets is that vessels that deploy drift gillnets will abandon them at sea in an effort to evade enforcement vessels (National Oceanic and Atmospheric Administration, 2008a). Derelict gillnets are found worldwide on beaches, reefs, and adrift at sea.

Ghost gillnets entangle fish, cephalopods, crustaceans, birds, turtles, marine mammals, vessels, and unwary humans (divers). The ghost fishing potential of gillnets varies considerably, depending primarily on the rigidity or permanence of the supporting mechanism(s). For example, pelagic drift gillnets typically hang between a cork line and a lead line and are set more or less in a line without fixed endpoints. Thus, they are subject to significant deformation by waves and currents. These nets have been shown to collapse over periods of days (Gerrodette et al., 1985), greatly reducing their long-term ghost fishing potential. Set gillnets, by virtue of their fixed, anchored framing, may remain fully deployed and fishing long after they are lost or abandoned (Carr et al., 1985). Ghost gill-nets in the shallow temperate waters of Puget Sound and in the Columbia River have been observed to self bait such that predators and scavengers attracted to entangled animals are themselves entangled, thereby per-petuating the cycle of destruction (Kappenman and Parker, 2007; Natural Resources Consultants, Inc., 2007). For example, a derelict net off Lopez Island in Puget Sound that had been in place for 15 years is estimated to have caught over 16,500 invertebrates, 2,340 fish, and 1,260 seabirds (Natural Resources Consultants, Inc., 2008). Additionally, one derelict gillnet, whose location was known for several years before a joint NOAA/ U.S. Coast Guard (USCG)-led multiagency effort was able to recover it in 1999, weighed over 2,000 kg (Donohue et al., 2001). As nets become fouled, they become more visible and lose their vertical profile, and their fishing capacity declines.

Traps, Cages, and Pots

Traps and pots are cages with wire, webbing, or other mesh, on a rigid or collapsible frame made of metal, wood, and other materials. They can range in size from small and light to quite large and heavy. Trap fisheries for crustaceans and finfish are carried out in relatively shallow, productive coastal and shelf areas worldwide. Fishing traps are typically weighted to sink and stay on the bottom with an attached marker float to allow reloca-tion and hauling back to the surface. Traps can be fished singly or con-nected together in strings. Traps are fitted with entry doors designed to prevent escape of the catch once inside. They are usually baited to attract the target species and retrieved and rebaited on a schedule suited to the catch rates, weather, and regulations of specific fisheries. Trap fisheries have also benefited from the replacement of natural fiber webbing and lines with more durable synthetic materials.

Trap loss has many causes, but includes weather-related movement and damage and loss due to conflicts with other user groups. Fishing vessels snag and move trap gear, floats are snagged on passing vessels or in towed fishing gear, competitors are reported to vandalize each other's gear in some fisheries (Acheson, 1977), and fishery closures and economic circumstances prevent or inhibit gear retrieval. Several trap fisheries in the United States are reported to have left tens or even hundreds of thousands of derelict traps in their fishing areas (Stevens et al., 2000; Guillory et al., 2001). In the Gulf of Mexico's spiny lobster and stone crab fisheries, traps are lost continually through the fishing season at a rate of 1–2 percent each month or 20–50 percent annually (Thomas Matthews, personal communication).

Traps catch legal and undersized target species as well as many other species that are attracted to the bait and are small enough to pass through the trap doors (Smolowitz, 1978; High, 1985; Breen, 1990). Traps often include holes sized to permit the escape of undersized target and nontarget species; bycatch unable to escape is typically released each time the trap is retrieved, but the loss of a trap means that animals that are unable to escape from the trap will starve or be preyed upon by their fellow captives. This catching and self-baiting cycle can continue for days to years until the trap is disabled, usually by being buried in sediments, general disintegration and biofouling, disintegration of degradable escape panels, if required and present, or by retrieval. Although most U.S. trap, cage, and pot fisheries require that they be equipped with rot cord (i.e., sections of twine that compromise the integrity of the pot once they biodegrade), Stevens et al. (2000) reports that one such equipped ghost pot alone held 125 crabs. An alternative to rot cord, galvanic releases—often used in pop-up oceanographic devices—offer greater consistency in the time to failure but have not yet won widespread application as release devices for fishing traps.

The loss of commercial species in derelict traps can be substantial. For example, derelict traps are estimated to account for about 7 percent of total mortality in the Dungeness crab fishery off the Fraser River delta in British Columbia (Breen, 1987). Matsuoka et al. (2005) shows that the take of octopus by derelict traps in a bay in Japan was equal to or twice that landed annually by the commercial fishery. Depending on the time of year, trap type, and location, annual ghost fishing mortality in the blue crab fishery ranges from 7.7 to 60 crabs per trap (Guillory et al., 2001). Marine mammals—in particular right whales, humpback whales, and dolphins—and sea turtles have been observed entangled in traps or buoy lines or groundlines (i.e., the line between traps) (National Oceanic and Atmospheric Administration, 2007a). While these entanglements are most likely to occur in active gear rather than in derelict gear, the extent to which derelict gear poses a threat to these marine mammals is unknown and indeed these entanglements are one vector for turning active gear into DFG.

Hook-and-Line

A wide variety of hook-and-line fisheries operates worldwide, including commercial longline, troll, jig, and dinglebar fishing, and recreational fishing using rod-and-reel or hand lines. These fisheries contribute to the DFG problem through the loss and discard of primarily monofilament fishing lines, although many other synthetic materials are used in braided fishing lines that may be equally persistent when derelict (e.g., Dacron®, Spectra®, and aramid polymers such as Kevlar®).

Loss of these lines is commonly caused by snagging and breakage during retrieval attempts, and by discard of snarled and damaged line. Longlines may be fished as drift or set (anchored) gear and can be lost due to weather, currents, damage by conflicting fishing activities (e.g., trawling) and other vessel traffic, as well as by vessel and equipment failures. Intense derby fisheries may result in longlines being set across each other, rendering the whole irretrievable (National Research Council, 1999).

The impact of derelict fishing lines is most obvious and dramatic when it entangles sea turtles, sea birds, and marine mammals; however, virtually all marine animals are susceptible to this entanglement (e.g., Shomura and Godfrey, 1990). Fishing line entanglement ordinarily results in traumatic amputation; strangulation; or other disablement leading to infection, starvation, heightened risk of predation, or death for the victim. Coral damage and death from entanglement in derelict monofilament fishing line (and associated sinkers and steel hooks) has been documented in South Africa (Schleyer and Tomalin, 2000), in the Mediterranean in northeastern Italy (Bavestrello et al., 1997), and in Hawaii (Asoh et al., 2004; Yoshikawa and Asoh, 2004).

Other Gear Types

Other major types of fishing gear include purse seines, shellfish dredges, FADs, shore-based fish traps and weirs, and net pens, cages, mesh bags, and lines used for aquaculture (coastal and offshore). These activities lose gear but, with the possible exception of FADs in the tropical tuna fisheries (see below), they are not yet documented as contributing to the overall impacts of DFG in the same magnitude as trawls, gillnets, traps, and line fisheries. Likewise, trammel nets, which are replacing gillnets in some fisheries, and lobster nets, which are replacing traps especially in coral reef habitats, while not yet documented, may also contribute to DFG. The growth of coastal and offshore aquaculture in Asia, Latin America, and Europe suggests that materials used in aquaculture are likely to become a more prominent component of DFG.

Legal and Regulatory Issues

There is some confusion among international and U.S. agencies over who is supposed to prevent, by regulation or otherwise, the generation of marine debris by fisheries, especially DFG because it is both a marine debris and a fishery management problem. The various international and domestic laws and regulations that govern activities that generate DFG and other marine debris from fisheries are summarized here.

While it is generally recognized that fishing is one of the freedoms of the seas, the discharge of unwanted fishing gear or the careless loss of useful or waste gear is not. Under the United Nations Convention on the Law of the Sea, "all states have the right of their nationals to engage in fishing on the high seas subject to their treaty obligations . . . [and to] the rights . . . and the interests of coastal states" (Article 116). Treaty obligations that are relevant to the DFG problem and its impacts on marine life include the environmental protection provisions of the Law of the Sea Convention (summarized in Chapter 3), the United Nations Fish Stocks Agreement of 1995, regional international fisheries agreements, and the provisions of MARPOL and its annexes. The conservation qualification of the high seas right of fishing is detailed in the United Nations Fish Stocks Agreement of 1995, which states that nations that fish for straddling and highly migratory fish species (which include tunas) have "a duty to adopt measures to minimize . . . catch by lost or abandoned gear" of both target and nontarget species through the development of environmentally safe fishing gear and techniques (Article 5[f]). Nations that are parties to both MARPOL Annex V and one or more international fishing agreements are summarized in Appendix D. Under international law, both coastal states and flag states (nations that register fishing vessels) bear responsibility to prevent marine debris, including DFG, by providing adequate reception facilities at fishing ports and enforcing regulations requiring proper disposal of waste fishing gear.

MARPOL Annex V

MARPOL Annex V addresses waste fishing gear in its ban on the discharge of plastics in all areas of the sea. Regulation 3 prohibits the disposal of "all plastics, including but not limited to synthetic ropes, synthetic fishing nets, [and] plastic garbage bags" (International Maritime Organization, 2006c), and a similar ban applies to special areas (Regulation 5). Regulation 6, however, exempts these discharges from the prohibition if they involve an "accidental loss of synthetic fishing nets, provided that all reasonable precautions have been taken to prevent such loss" (International Maritime Organization, 2006c). Therefore, it is not a violation if plastic or other synthetic fishing gear falls overboard due to damage to the fishing vessel or its equipment, provided that all reasonable precautions were taken to prevent such loss, or if the gear is intentionally put overboard in order to secure the safety of the vessel, its crew, or lives at sea.

The International Maritime Organization (IMO) amended the accidental loss exception in 1989 to exclude the loss of pieces or fragments of nets from on deck, presumably because member states concluded that there are no reasons to excuse the loss of such items (Koehler et al., 2000). Thus, any time a synthetic fishing net or piece of net or another type of plastic fishing gear is thrown overboard for deliberate disposal or purposefully left behind, that action constitutes a prohibited disposal under MARPOL Annex V.

IMO's member states, acting through IMO's Marine Environmental Protection Committee (MEPC), recognize that compliance with MARPOL Annex V has been incomplete and that marine debris, including debris from fishing operations, is a substantial threat. In 2006, MEPC charged a group of member states, coordinated by Canada, to complete a review of MARPOL Annex V and its *Guidelines* (International Maritime Organization, 2007; see also Chapter 3). While the review by the MEPC correspondence group is ongoing as of this report, a number of fisheries-related problems have been identified and options to rectify them have been suggested by members of the correspondence group (International Maritime Organization, 2007). Here some of these points are summarized, and it is noted that the committee's analysis is very much in line with these findings:

- Regulation 6(c) of MARPOL Annex V contains no definition of "reasonable precautions" for the exception allowing the "accidental loss of synthetic fishing nets, provided that all reasonable precautions have been taken to prevent such loss" (International Maritime Organization, 2006c). Without a definition, it is very difficult to enforce. The regulations could better define "reasonable precautions," although there are many challenges to doing so.
- Regulations 9(2) and 9(3)(d) (International Maritime Organization, 2006c) only require fishing vessels over 400 gross tons to maintain a garbage management plan and report the loss of fishing gear in either a garbage record book or a special reporting system designed for fishing gear. Because only a small number of fishing vessels are this large, very few vessels are required to report lost fishing gear, even though this would help authorities to retrieve it, analyze its impact, or mitigate its potential damage. A provision could be added that requires all fishing vessels (except artisanal and noncommercial vessels) to maintain garbage management plans and record losses of gear.
- MARPOL Annex V and its regulations do not include a requirement to mark gear so that it can be identified and traced to its source. If this were included, it would likely increase the number

of gear marking requirements that fisheries management organizations adopt and which flag nations require. The Food and Agriculture Organization of the United Nations (FAO) (1995) has stated that "fishing gear should be marked in accordance with national legislation in order that the owner of the gear can be identified. Gear-marking requirements should take into account uniform and internationally recognizable gear marking systems." In Section 3.5.3, the *Guidelines for the Implementation of Annex V of MARPOL* (International Maritime Organization, 2006b; see also Box 4.1) encourages the use of gear identification systems, but this does not seem to be a widespread practice. A gear-marking provision could be added to MARPOL Annex V.

- The *Guidelines* (International Maritime Organization, 2006b) contains a great deal of information related to fishing gear that does not appear to be widely applied; this information could be made more accessible to fisheries management organizations and stakeholders.
- The phenomenon of illegal, unreported, and unregulated fishing in the world's oceans is also relevant to the work of the correspondence group. A joint IMO/FAO ad hoc working group on illegal, unreported, and unregulated fishing has been formed and this group has requested advice from FAO and other United Nations agencies on measures that IMO could adopt for fishing vessels that would help to combat this problem.

IMO's *Guidelines for the Implementation of Annex V of MARPOL* (International Maritime Organization, 2006b) was published in 1988 and revised in 1991. It contains much of the detail that is lacking in the regulations (see Box 4.1), perhaps because IMO parties assumed that member states needed flexibility in implementing MARPOL Annex V and would be more likely to comply with Annex V if they had guidance rather than detailed rules. The correspondence group discussions indicate that the time may have come to put more specificity into the regulations with respect to fishing gear (International Maritime Organization, 2007).

Finding: MARPOL Annex V does not adequately or comprehensively manage discharges of fishing gear into the marine environment. The exemption for "the accidental loss of synthetic fishing nets, provided that all reasonable precautions have been taken to prevent such loss" (International Maritime Organization, 2006c), does not provide sufficient guidance to regulators and the fishing industry. Moreover, because of minimum length and gross tonnage exemptions, MARPOL Annex V does not apply to a substantial number of commercial, artisanal, and sport fishing charter vessels. Therefore, these vessels

BOX 4.1
*Guidelines for the Implementation of Annex V of MARPOL
for Fishing Gear*

3.5. Fishing gear, once discharged, becomes a harmful substance. Fishing vessel operators, their organizations and their respective governments are encouraged to undertake such research, technology development and regulations as may be necessary to minimize the probability of loss, and maximize the probability of recovery of fishing gear from the ocean. It is recommended that fishing vessel operators record and report the loss and recovery of fishing gear. Techniques both to minimize the amount of fishing gear lost in the ocean and to maximize recovery of the same are listed below.

3.5.1. Operators and associations of fishing vessels using untended, fixed or drifting gear are encouraged to develop information exchanges with such other ship traffic as may be necessary to minimize accidental encounters between ships and gear. Governments are encouraged to assist in the development of information systems where necessary.

3.5.2. Fishery managers are encouraged to consider the probability of encounters between ship traffic and fishing gear when establishing seasons, areas and gear-type regulations.

3.5.3. Fishery managers, fishing vessel operators and associations are encouraged to utilize gear identification systems which provide information such as vessel name, registration number and nationality, etc. Such systems may be useful to promote reporting, recovery and return of lost gear.

3.5.4. Fishing vessel operators are encouraged to document positions and reasons for loss of their gear. To reduce the potential of entanglement and "ghost fishing" (capture of marine life by discharged fishing gear), benthic traps, trawl and gillnets could be designed to have degradable panels or sections made of natural fiber twine, wood or wire.

3.5.5. Governments are encouraged to consider the development of technology for more effective fishing gear identification systems.

7.1.4. Governments should consider . . . the use of garbage discharge reporting systems (e.g., existing ship's deck logbook or record book) for ships . . . [to] document the date, time, location by latitude and longitude, or name of port, type of garbage . . . and estimated amount of garbage discharged. . . . Particular attention should be given to the reporting of . . . the loss of fishing gear . . .

SOURCE: International Maritime Organization (2006b).

are exempt from placarding, garbage management plan, and garbage log requirements that would facilitate enforcement of prohibitions against the at-sea disposal of synthetic fishing gear.

Recommendation: The U.S. delegation should exercise its influence in the correspondence group and on IMO's MEPC to amend MARPOL Annex V to

- provide explicit definitions of "accidental losses" and "reasonable precautions" with respect to synthetic fishing nets;
- require placards, garbage management plans, and record books for all commercial, artisanal, and sport fishing charter vessels to the extent practicable, recognizing that some exceptions, perhaps by vessel size or gear type, will be necessary; and
- require additional practices that minimize the probability of loss and maximize the probability of recovery of fishing gear from the ocean, including (1) development of improved information systems and fisheries management measures that reduce conflicts between fishing gear and other user groups, (2) requirements for gear marking and identification systems, (3) documentation of position and reasons for gear loss, and (4) inclusion of degradable elements in synthetic gear to reduce the potential of entanglement and ghost fishing.

Domestic Implementation of MARPOL Annex V for Derelict Fishing Gear

To implement MARPOL and Annexes I and II, Congress enacted the Act to Prevent Pollution from Ships (APPS) (33 U.S.C. § 1901 et seq.) in 1980, giving USCG regulatory power to carry out their provisions for U.S. flag vessels, and foreign vessels in U.S. waters and ports. When the United States then ratified MARPOL Annex V seven years later, Congress enacted the Marine Plastic Pollution Research and Control Act of 1987 (MPPRCA) (33 U.S.C. § 1901 et seq.) to provide for its implementation in U.S. waters and by U.S. vessels. MPPRCA amended APPS to require USCG to adopt regulations to implement MARPOL Annex V's ban on the disposal of synthetic fishing nets and other fishing-related plastic garbage and its requirement that vessels maintain refuse record books, garbage management plans, and post placards summarizing MARPOL Annex V (see Chapter 3). Notably, Congress put in provisions that would ultimately require USCG to apply the requirements of MARPOL Annex V to all ships calling at U.S. ports, regardless of their state of registry (i.e., their "flag state").

USCG promulgated final regulations to implement MPPRCA in 1990, and it later revised these regulations to incorporate IMO's amendments to MARPOL Annex V (33 C.F.R. §§ 151.51–151.77). USCG regulations require that all manned oceangoing vessels 400 gross tons and larger, fixed and floating platforms, and every vessel certified to carry more than 15 passengers engaged in international voyages keep records of their garbage discharges and disposal (including accidental) (33 C.F.R. § 151.55) and that all vessels of 12.192 meters (40 feet) or more in length and fixed or floating platforms have waste management plans (33 C.F.R. § 151.57). In addition, the master of a ship is required to notify the port upon its approach of the estimated volume of garbage it has onboard and whether any of it requires special handling, which would presumably include a large fishing net that has outlived its usefulness. Tracking MARPOL Annex V Regulation 3, the U.S. regulations state that "no person on board any ship may discharge into the sea . . . plastic or garbage mixed with plastic, including, but not limited to, synthetic ropes, synthetic fishing nets, and plastic garbage bags" (33 C.F.R. § 151.67). Like MARPOL Annex V, this prohibition is subject to an emergency exception for accidental fishing gear losses if all reasonable precautions have been taken. However, the regulations do not define what the standard is for these reasonable precautions, making the prohibition in Section 151.67 very difficult to enforce. This may explain why there is no evidence that USCG has taken any MARPOL Annex V enforcement actions against fishing vessels for their failure to prevent accidental losses of fishing gear (e.g., U.S. Coast Guard, 2007).

The requirements imposed by the regulations thus constitute the minimum necessary to implement MARPOL Annex V. One explanation for the lack of detail regarding fishing vessels and waste fishing gear may lie in the narrow manner in which USCG interprets its regulatory powers under MPPRCA. MPPRCA gives USCG broad regulatory powers to give effect to MARPOL Annex V's ban on disposal of plastic, including synthetic fishing gear, at sea (33 U.S.C. § 1902–1903). However, USCG has eschewed adopting measures that would help to prevent the generation of DFG and has deferred consideration of measures like gear identification systems or programs to encourage recovery of lost or discarded fishing gear to other agencies.[2]

USCG expressed this interpretation when it adopted the final rules to implement MPPRCA in 1990. It stated that Congress had addressed the need to control the impact of driftnets on the marine environment in Title IV of MPPRCA and in the Driftnet Impact Monitoring, Assessment, and Control Act of 1987 (16 U.S.C. § 1822 et seq.) (Box 4.2) and required the Secretary of Commerce, not USCG, to evaluate the need for systems of marking, registering, and identifying driftnets and for paying bounties

---

[2]For example, in the preamble to the 1989 interim final rules, USCG described its authority in the following response to public comments:

> Two commenters requested USCG to initiate a fishing gear marking, registration and identification system and to develop a monetary bounty system to aid in the recovery of fishing gear. USCG has determined it does not have the authority to accomplish either of these under Title II of Pub. L. 100-220. As discussed above, however, USCG is actively considering what recordkeeping requirements would be appropriate for commercial fishing vessels to achieve the goals of Annex V. (54 Fed. Reg. 18384–18389 [April 28, 1989]).

BOX 4.2
The Driftnet Impact Monitoring, Assessment, and Control Act and Derelict Fishing Gear

When the Marine Plastic Pollution Research and Control Act was enacted in 1987, Congress' main concern with respect to waste fishing gear was in connection with the rapidly expanding large-scale driftnet fisheries on the high seas. The public law containing the Marine Plastic Pollution Research and Control Act also included the Driftnet Impact Monitoring, Assessment, and Control Act, which called for agreements with high seas fishing nations to study the extent and impacts of driftnet fishing in the North Pacific on marine resources of the United States and to control the location, season, and other aspects of such fishing to prevent those impacts. If foreign nations failed to enter these agreements, the U.S. Secretary of Commerce was directed to certify that fact to the President, who would in turn consider imposing trade sanctions under the Pelly Amendment (22 U.S.C. § 1978).

Congress also directed the Secretary of Commerce to research the feasibility "of a driftnet marking, registry, and identification system to provide a reliable method for the determination of the origin by vessel of lost, discarded, or abandoned driftnets and fragments of driftnets" (16 U.S.C. § 1822 et seq.) and to evaluate the adequacy of existing identification systems used in foreign driftnet fisheries. The Secretary was also directed to develop recommendations that would require driftnets to be made of material that would more readily decompose if discarded or lost at sea, a driftnet bounty system, and a driftnet vessel tracking system, and to report all these things to Congress. The bounty system was "to pay persons who retrieve from the exclusive economic zone and deposit with the Secretary lost, abandoned, and discarded driftnet and other plastic fishing material" (16 U.S.C. § 1822 et seq.).

to fishing or merchant vessels that retrieve driftnets. It then concluded, "[n]either these evaluations nor any resultant legislative or regulatory action falls under the responsibility of the Coast Guard" (55 Fed. Reg. 35986–35987 [September 4, 1990]). USCG appears by this response to read MPPRCA to authorize only NOAA to consider federal programs for other types of fishing gear, acting under the Driftnet Impact Monitoring, Assessment, and Control Act, the Magnuson-Stevens Fishery Conservation and Management Act of 2006 (MSFCMA) (16 U.S.C. § 1801 et seq.), or some other fisheries legislation.

Given these responses, it appears that USCG is unlikely to adopt additional DFG-related regulations without a direct congressional mandate and budget authorization, even though APPS as amended gives USCG broad rulemaking authority to implement items recommended in the *Guidelines* (International Maritime Organization, 2006b) for preventing DFG and to adopt measures to define ambiguous terms in MARPOL Annex V and APPS. However, the latest enactment on plastic pollution by Congress—the Marine Debris Research, Prevention, and Reduction

Act of 2006—does not direct USCG to undertake a rulemaking to determine, inter alia, which of the MARPOL Annex V *Guidelines* (International Maritime Organization, 2006b) would be appropriate and effective in reducing DFG generated by U.S. fishing vessels and fisheries in U.S. waters. Because the 2006 Act reestablishes the NOAA Marine Debris Prevention and Removal Program and requires the NOAA Administrator to undertake marine debris identification, prevention, and removal efforts, USCG is likely to continue to defer to NOAA to adopt DFG prevention measures. However, Congress has made implementation of MARPOL Annex V in the United States the responsibility of the Secretary of the department in which USCG is operating—the Department of Homeland Security—under the terms of the Driftnet Impact Monitoring, Assessment, and Control Act. USCG also has responsibility to make sure U.S. ports and waterways are free of navigational hazards under the Ports and Waterways Safety Act (33 U.S.C. § 1221 et seq.). It is the duty of the Secretary of Commerce to ensure that fisheries are sustainable and to protect the marine environment from any adverse effects of fishing. Given these overlapping authorities, a program of joint rulemaking may be in order. Congress may, however, need to require this expressly, as it did with respect to rerouting oil tanker traffic to protect national marine sanctuaries off the California coast, which led in turn to a successful proposal to IMO for revised international ship routing measures (National Marine Sanctuaries Act of 1972, 16 U.S.C. § 1431 et. seq.).

Finding: Both USCG and NOAA have rulemaking authority to prevent the generation of DFG under their respective legislative mandates, yet neither has exercised that authority to require the adoption of measures to prevent the loss of fishing gear, to document the location of unpreventable losses of fishing gear, or to encourage proper disposal of waste fishing gear.

Recommendation: Congress should direct USCG and NOAA to undertake a joint rulemaking to develop rules that require commercial and recreational fishing vessels to properly dispose of all waste fishing gear and to take specific precautions to prevent accidental loss of fishing gear.

International Fisheries Agreements

Prevention and reduction of DFG is clearly a part of sustainable and responsible fisheries management; therefore, international fisheries agreements play a role in preventing and reducing this type of marine debris. This includes commitments made by nations to manage fisheries under Regional Fisheries Management Organizations (RFMOs), such as the Inter-American Tropical Tuna Commission (IATTC) and the Western and Central Pacific Fisheries Commission (WCPFC). These commitments could include measures to reduce the loss of fishing gear. In addition, many nations with high seas fishing fleets have treaty obligations under the United Nations Fish Stocks Agreement of 1995 that require them to cooperate with other nations to ensure that fish stocks and other resources of the marine environment are not endangered by fishing. To give effect to their duty to cooperate, coastal states and states fishing on the high seas must, inter alia, "minimize pollution, waste, discards, catch by lost or abandoned gear, catch of non-target species, . . . and impacts on associated and dependent species, in particular endangered species, through measures including, to the extent practicable, the development and use of selective, environmentally safe, and cost-effective fishing gear and techniques" (United Nations Fish Stocks Agreement, Article 5).

The member states of FAO have also adopted an international "Code of Conduct for Responsible Fisheries" (Food and Agriculture Organization of the United Nations, 1995). The Code represents a commitment by nations to work through relevant organizations at national and international levels to ensure that vessels that fly their flags fish responsibly. The Code's overall commitment is that "the harvesting, handling, processing, and distribution of fish and fishery products should be carried out in a manner which will . . . minimize negative impacts on the environment" (Food and Agriculture Organization of the United Nations, 1995). On the issue of DFG, the Code provides that states, RFMOs, and subregional bodies should adopt appropriate measures to minimize catch by lost or abandoned fishing gear and its impact on nontarget species, in particular endangered species (Food and Agriculture Organization of the United Nations, 1995). Also, fishing activities should be conducted with due regard to the IMO requirements relating to the protection of the marine environment and the loss of fishing gear (Food and Agriculture Organization of the United Nations, 1995; Koehler et al., 2000).

Unfortunately, despite the obligations of the United Nations Fish Stocks Agreement of 1995 and the closely related FAO Code, few if any international fishery organizations have taken steps to prevent damage to the marine environment from DFG. The international fishery organizations are struggling even to meet their core legal obligations to adopt necessary conservation and management measures to prevent overexploitation and to allocate equitably the burdens of those measures among high seas fishing states and coastal states. With these difficulties, marine debris is at best a third-order priority.

One exception is the Commission for the Conservation of Antarctic Marine Living Resources (CCAMLR), which has an active program to combat marine debris, including debris from fishing activities such as large-scale trawl fisheries for krill and longline fishing for Patagonian toothfish. CCAMLR requires that all fishing and fishing research vessels have identifying marks on each item of fishing gear, post marine debris placards that include a symbol prohibiting fishing net disposal, carry observers, and distribute educational materials explaining the marine debris regulations in force. Monitoring marine debris and its impacts is a permanent agenda item of CCAMLR and its scientific committee. Members submit yearly surveys of debris on beaches and in seabird colonies, of marine wildlife entanglements, and of hydrocarbon soiling of mammals and seabirds. The Secretariat maintains a marine debris database from 12 index sites on the Antarctic Peninsula and on Antarctic and sub-Antarctic islands.

CCAMLR's active interest in addressing the marine debris—including DFG—problem may be due in part to the designation of the waters south of 60°S latitude as a MARPOL Annex V special area in 1992. CCAMLR is also included in the Antarctic Treaty System. The treaty system has an environmental protocol and annex that includes measures to prevent marine pollution and tracks MARPOL provisions closely. Also, the mandate of CCAMLR includes the principle that harvesting activities must minimize the risk to the Antarctic marine ecosystem (Commission for the Conservation of Antarctic Marine Living Resources, Articles II and IX). In 2006, CCAMLR adopted a binding conservation measure prohibiting the use of plastic packaging bands on fishing vessels which do not have onboard incinerators and requiring the cutting of bands on those vessels that do. Box 4.3 summarizes CCAMLR's measures to combat marine debris and DFG.

BOX 4.3
Commission for the Conservation of Antarctic Marine Living Resources Legally Binding Measures to Reduce Marine Debris

Conservation Measure 10-01: Marking of Fishing Gear—all fishing gear such as pots, marker buoys, floats, etc. must be marked with the vessel name, call sign and flag state

Conservation Measure 25-01: Regulation of the Use and Disposal of Plastic Packaging Bands on Fishing Vessels

Conservation Measure 25-02: Minimization of the Incidental Mortality of Seabirds in the Course of Longline Fishing or Longline Research in the Convention Area

SOURCE: Appleyard (2004).

There are many cost-effective and technically feasible measures that could be adopted by RFMOs that would reduce the accumulation of damaging DFG. These include the following:

- binding conservation measures to prohibit discarding fishing gear, light sticks, plastic packaging and lines, and all other plastic materials used onboard fishing vessels;
- requirements to carry a certified manifest of fishing gear onboard at the beginning of a trip to allow port states to inspect for compliance with MARPOL Annex V and refer violations for flag state enforcement;
- measures requiring vessels to mark individual pieces of fishing gear and to log and call in timely reports of accidental or willful discharges of fishing gear;
- solid waste management plans that specifically address waste fishing gear;
- observer reports of discharged fishing gear;
- bounty systems, deposits, or other incentives for the retrieval of DFG encountered at sea (including gear collected in trawl nets) for recycling and proper disposal; and
- rewards or regulatory priority given to fishing vessels that report and record waste fishing gear discharges that are successfully prosecuted.

Some of these measures are new but many are detailed in the MARPOL Annex V *Guidelines* (International Maritime Organization, 2006b; see Box 4.1).

Another factor contributing to the amount of DFG is likely to be the rapidly growing phenomenon of illegal, unreported, and unregulated fishing in the world's oceans (Kock, 2001; Ilse Kiessling, personal communication). To avoid seizure, vessels engaged in illegal high seas fisheries have been reported to cut their nets free and to flee from approaching enforcement vessels (e.g., drift gillnets; National Oceanic and Atmospheric Administration, 2008a). At the joint IMO/FAO ad hoc working group on illegal, unreported, and unregulated fishing meeting in 2007, the coordinator of the IMO correspondence group requested advice from FAO and other United Nations agencies on measures that IMO could adopt for fishing vessels that would help to combat this problem. The correspondence group also noted the adoption of a memorandum of understanding between FAO and the United Nations Environment Programme to undertake a study on the issue of abandoned, lost, or otherwise discarded fishing gear, and that FAO agreed to develop standards for the marking of fishing gear and the location and retrieval of lost fishing gear through technologies such as barcoding and transponders (International Maritime Organization, 2007).

Comprehensive management of fishing gear use and disposal will be most effective if improvements to MARPOL Annex V are also accompanied by parallel RFMO actions. As previously discussed, amendments to MARPOL Annex V that contain more detailed provisions on fishing gear would greatly assist the international and regional bodies that seek to manage fisheries and to protect the marine environment. If these international conventions adopted comparable DFG prevention measures, it would ensure implementation of these measures by additional fishing fleets that may not be signatories to each individual convention. For example, one of the major high seas fishing fleets is flagged by a political entity—Taiwan—that is not eligible for membership in IMO. Taiwan is, however, a participating entity in several international fisheries organizations and would be bound to comply with measures adopted by those organizations related to waste fishing gear. (See Appendix D for a table of parties to MARPOL Annex V and various RFMOs.)

U.S. Fisheries Management

U.S. marine fisheries are managed under the authority of federal and state laws, as conditioned by treaties with sovereign tribes and foreign nations. In general, state authority extends 3 miles from shore; in Texas and on the west coast of Florida, state waters extend to 9 miles offshore. Multistate compacts coordinate state management of migratory stocks of fish and shellfish. While states have management autonomy within state waters, state management cannot impinge on management of federal resources (Bader, 1998). In practice, this means that there is a need for coordination between state and federal fishery managers except in unusual circumstances where a resource only occurs in state waters or only in federal waters. Coordination is often accomplished by the passage of parallel management measures in federal and state waters or by deference in one direction or the other. In the case of crab fisheries off Alaska, for example, the requirement for gear marking in the federal fishery derives from a requirement imposed in the state fishery and from federal deference to the state for management of the crab fisheries in state and federal waters. Because federal authority to enact treaties is superior to state authority, ratification of MARPOL Annex V compels states to ensure that their fishery management practices are consistent with MARPOL Annex V requirements.

Just as it is legal for individuals to buy, use, and dispose of a very wide range of persistent synthetic products, it is also legal to participate in fisheries using gear and equipment that, by their nature, can become potentially hazardous forms of marine debris. Ideally, these activities are conducted in a regulated environment to ensure that the intended usage and impacts are controlled in the interests of society. Under MSFCMA, NOAA adopted regulations that subject foreign fishing that takes place in U.S. waters to a series of conditions and requirements aimed at combating the DFG problem (50 C.F.R. § 600.510(c)(1)–(3); Box 4.4). While the elimination of foreign fishing within the U.S. EEZ (e.g., National Oceanic and Atmospheric Administration, 2007b) has rendered these measures moot, they provide a template for possible measures directed at the domestic fleet, which is not currently subject to measures that require minimizing the loss, maximizing the recovery of, or limiting the hazards presented

---

BOX 4.4
Magnuson-Stevens Fishery Conservation and Management Act Measures to Combat Derelict Fishing Gear from Foreign Fisheries in the U.S. Exclusive Economic Zone

The National Oceanic and Atmospheric Administration (NOAA) Fisheries Service has adopted a set of federal regulations that apply to any and all foreign fisheries should they be authorized in the U.S. Exclusive Economic Zone (EEZ). In addition to requiring foreign vessels to have permits, onboard observers, and recordkeeping and to facilitate enforcement, the regulations contain an express prohibition on the disposal or abandonment of fishing gear. Foreign fishing vessels are also required to report accidental loss or emergency jettisoning of gear to the U.S. Coast Guard (USCG). No similar blanket DFG regulation exists for domestic commercial and sport fishing vessels.

NOAA's regulations for gear avoidance and disposal outline the following parameters regarding the disposal of fishing gear and other items:

(1) The operator of a [foreign fishing vessel] in the EEZ may not dump overboard, jettison, or otherwise discard any article or substance that may interfere with other fishing vessels or gear, or that may catch fish or cause damage to any marine resource, including marine mammals and birds, except in cases of emergency involving the safety of the ship or crew, or as specifically authorized by communication from the appropriate USCG Commander or other authorized officer. These articles and substances include, but are not limited to, fishing gear, net scraps, bale straps, plastic bags, oil drums, petroleum containers, oil, toxic chemicals, or any man-made items retrieved in a [foreign fishing vessel's] gear.
(2) The operator of a [foreign fishing vessel] may not abandon fishing gear in the EEZ.
(3) If these articles or substances are encountered, or in the event of accidental or emergency placement into the EEZ, the vessel operator must immediately report the incident to the appropriate USCG Commander . . .

(50 C.F.R. § 600.510(c)(1)–(3)).

by gear lost in those fisheries. Although MSFCMA (16 U.S.C. § 1801 § 206 (b)(3)) identifies a need for reliable estimates of the numbers of seabirds, sea turtles, nontarget fish, and marine mammals entangled and killed in derelict large-scale driftnets (i.e., high seas drift gillnets), it does not contain any other provisions that directly address the minimization, disposal, or removal of DFG or to prevent harm to wildlife populations or damage to sensitive marine ecosystems caused by DFG.

MSFCMA does, however, indirectly require regional fishery management councils (FMCs) to minimize DFG. National Standard 9 of the Act requires councils to devise conservation and management measures for fisheries within their region that "to the extent practicable, (A) minimize bycatch and (B) to the extent bycatch cannot be avoided, minimize the mortality of such bycatch" (16 U.S.C. § 1801 et seq.). Since ghost fishing by DFG can result in bycatch (Debra Lambert, personal communication), the councils could implement measures to ensure that DFG is minimized, including incentives for vessels to retrieve DFG encountered on the fishing grounds. The councils are also authorized to develop fishery regulations that designate zones where fishing shall be limited, not permitted, or permitted only to specified types of fishing vessels or gear. These zones may be designated to prevent loss or damage to fishing gear from interactions with deep sea corals, for example (Debra Lambert, personal communication). NOAA could also improve the understanding of gear loss and promote gear recovery through the use of observer reporting. Although many U.S. fisheries do not require onboard observers, some of the largest fisheries do. The primary purpose of these observers is to monitor and sample catches of target and incidental species and document interactions with marine mammals and other protected species; there is no requirement for observers to document gear loss at this time. However, this is something that observers could include in their recordkeeping, as long as this did not take priority over the observers' primary responsibilities for monitoring catch and discards, and any observer documentation is done in addition to (not in lieu of) reports made by vessel operators.

Information provided to the committee by the Caribbean Fishery Management Council, the New England Fishery Management Council, the North Pacific Fishery Management Council, the Gulf of Mexico Fishery Management Council, and the Western Pacific Fishery Management Council indicate that regional FMCs have included provisions in their fishery management plans (FMPs) that have direct or incidental effects on the quantity of DFG generated and the likelihood that lost or abandoned fishing gear will ghost fish. These provisions generally center around four themes: reducing the amount of gear, minimizing gear loss, minimizing ghost fishing and other impacts of fishing traps, and marking gear.

Reducing the Amount of Gear

One of the benefits often anticipated from the adoption of effort control measures, vessel buyback plans, and limited access privilege programs is a reduction in the amount of gear lost or abandoned. This anticipated benefit is thought to arise from a reduction in the amount of gear deployed—less gear fished, less gear lost (National Research Council, 1999).

Minimizing Gear Loss

Another benefit of effort control measures is a reduction in the amount of gear abandoned or lost due to multiple units of gear being fished in overly close proximity during compressed seasons or during inclement sea conditions encountered during short derby fishing openings (National Research Council, 1999). Measures that reduce conflicts among user groups (either between different fishing groups or between fisheries and other maritime sectors) generally reduce the frequency that stationary and mobile fishing gear become tangled and lost or abandoned. In addition, FMCs indicate that the primary motivations for closing high-profile substrate areas (e.g., coral reefs, seamounts, deep sea coral beds)—as marine protected areas or as gear exclosures—are to reduce the impact of gear on the substrate and to reduce gear losses. Similarly, management measures adopted to address the provisions of the Sustainable Fisheries Act (16 U.S.C. § 1801 et seq.) for essential fish habitat and habitat of particular concern have considered potential damage to living substrates and damage or loss of fishing gear. For example, the Caribbean Fishery Management Council has banned gillnets and trammel nets, in part out of concern that these gear types are prone to become entangled on reefs and that they are likely to damage living substrate during retrieval or to be abandoned as unrecoverable. The North Pacific Fishery Management Council established a requirement that crab pots fished in the eastern Aleutian Islands golden king crab fishery must be fished in connected strings of at least 10 pots. The purpose of this requirement was to reduce the loss of gear in an area with narrow ledges, steep bottom slopes, and strong currents.

Minimizing the Impact of Lost Gear

FMPs and conforming state and multistate fishery agency regulations for traps, cages, and pots in crab, lobster, and finfish fisheries generally require that traps include at least one escape panel that is secured with a cotton twine that will disintegrate within a few weeks or months, thereby reducing ghost fishing. However, as noted in Chapter 2, Stevens et al.

(2000) reported that one ghost pot equipped with rot cord off Kodiak was observed with 125 crabs, and Barnard (2008) determined that the mean failure rate for standard 30-thread cotton twine is 77–89 days; thus, even properly equipped traps could continue to ghost fish for an extend period. Galvanic releases might provide more consistent disintegration rates.

Gear Marking

There are two aspects of gear marking. Gear marking is used extensively in fisheries that employ static gear, primarily as a means of discouraging theft of catches and gear, to encourage postseason recovery of gear, and as a mechanism for enforcing individual limits on the amount of gear that may be deployed. This aspect of gear marking is not controversial. The other aspect of gear marking is focused on tracing DFG back to particular fisheries and fishing vessels; this aspect of gear marking is controversial. Through the mid- to late 1980s, there was a vigorous debate over the liability of fishermen for the damages caused by DFG. The premise was that if one could identify the owner of a specific piece of derelict gear through forensic analysis, there would be the potential to punish or seek other remedy from the offending party. Under those circumstances, any number of corrective or preventative actions might be implemented by exploiting this legal leverage. While gear marking poses implementation challenges (Henderson and Steiner, 2000), it is feasible and could serve as an important tool for understanding the dynamics of the DFG problem in fisheries subject to marking requirements. Effective gear marking is critical for identification of the sources of DFG and the fisheries that may have deployed this gear. Better information on loss rates and fates and effects of lost gear in the context of a broad, "no fault" DFG accountability and management regime will help focus technological innovation and recovery efforts on the highest priority sources. Any "no fault" provision would, however, require that losses and their circumstances be reported in a timely manner. Data from gear marking programs can also be used to inform outreach programs designed to motivate the involved community to be more responsive to DFG issues and solutions. Clearly, while marking of fishing gear is a valuable tool for fishery management and enforcement, it is unlikely to be a suitable means for deriving actionable evidence linking individual fishermen and the impacts of their lost gear.

Fisheries that adversely affect endangered marine species or their critical habitat are sometimes subject to regulations in addition to those implementing an FMP. Some of these additional rules, adopted under the Endangered Species Act (ESA) (16 U.S.C. § 1531 et seq.), are aimed at reducing the direct impacts of fishing gear as it is being actively fished, such as the accidental trapping or entanglement of a sea turtle (National Research Council, 1990) or at reducing the potential for localized competition between the fishery and the listed species (National Research Council, 2003). It is conceivable, however, that if the rate of gear loss or intentional discharge is high in a particular fishery, or if the accumulation of DFG in a critical habitat is particularly severe, causing injury or death to listed species, federal or state authorities could be required to adopt measures, such as gear marking, aimed at reducing the incidence of DFG. For example, in *Strahan v. Coxe* (127 F.3d 155 [1st Cir. 1997]), it was found that the state fisheries agency had third-party liability for causing takes of endangered whales by licensing use of fixed fishing gear in the whale habitat.

It is worth noting that regional FMCs, like the international fisheries organizations, may be reticent to adopt regulations that are perceived to increase costs or decrease catches or catch-per-unit effort. Nevertheless, such measures may be necessary and have been previously implemented to protect marine wildlife and the marine environment. NOAA would be within its authority to adopt generic fishery regulations requiring full life-cycle accountability for the deployment and retrieval of fishing gear out of concern for living marine resources or the marine environment. NOAA could do so under MSFCMA, ESA, or the Marine Mammal Protection Act (MMPA) (16 U.S.C. § 1361 et seq.). Regulations under ESA and MMPA would be applicable to fisheries in state as well as federal waters and thus could have a broader effect than actions taken solely under authority of MSFCMA.

Finding: MSFCMA does not highlight the need to reduce DFG or other marine debris nor does it contain a national standard to address DFG or other marine debris.

Recommendation: Congress should add a national standard to MSFCMA that fishery conservation and management measures shall be designed to minimize the risk of gear loss.

Finding: Although some FMPs currently include measures that may have a collateral benefit of reducing DFG, current FMPs do not include measures that specifically address DFG.

Recommendation: NOAA should establish a timetable for review of all existing FMPs for opportunities to reduce fishing-related marine debris, including reducing gear, minimizing gear loss, and minimizing impacts of lost gear, and to improve gear marking and recovery. Measures that reduce the loss or abandonment of fishing gear and encourage the retrieval of DFG should be considered in all future FMPs, National Environmental Policy Act (42 U.S.C. § 4321 et seq.)

documents, and ESA Section 7 consultations and biological opinions. NOAA should encourage adoption of these measures by fisheries management organizations at the regional, state, and international levels. NOAA should also expand the duties of observers to include documentation of gear loss.

Finding: Prevention of DFG begins at the source, but identifying the source may be difficult because ocean currents can transport DFG a long distance from the site of loss or discard and can involve substantial time lags. Effective gear marking is critical for identification of the sources of DFG and the fisheries that may have deployed this gear.

Recommendation: NOAA should convene a workshop to explore innovative and cost-effective approaches for identification or marking of trawls, seines, gillnets, longlines, and FADs to foster gear identification. Based on this information, NOAA should develop gear marking protocols that can be used in domestic and international fisheries to provide a structured basis for designing programs to reduce gear loss and abandonment and increase recovery of DFG.

Finding: DFG has the potential to negatively impact endangered and protected species. For those fisheries that generate DFG that harms endangered and protected species, NOAA has the authority under ESA and MMPA to require fishing gear accountability measures.

Recommendation: NOAA should

- determine which endangered and protected marine wildlife species or populations are at risk in part from DFG based on a review of all available information on fisheries interactions with these species;
- include information on injury and deaths due to DFG or other fishing-related marine debris in its marine mammal stock assessments and recovery plans and status reports for other threatened and endangered species; and
- use the provisions of ESA and MMPA to require adoption of gear accountability and other measures to minimize or remove DFG for fisheries that generate DFG that poses an entanglement threat to endangered and protected marine wildlife.

Challenges with Gear Recovery and Disposal

While there are many admirable examples of DFG recovery programs in the United States and beyond, there are many challenges to gear disposal and no clear responsibility for derelict gear recovery. Adequate and affordable disposal of fishing gear can be problematic as the gear is often very bulky and fishing vessels often operate in remote and less populated areas with limited waste management capacity. Similarly, there are many regulatory and practical challenges with management of recovered derelict gear, not only in finding adequate disposal facilities but also because of legal hurdles that may discourage the recovery of gear from the environment.

Responsibility for Gear Recovery

Currently, recovery of DFG is undertaken by a variety of groups, particularly government agencies at various levels and nongovernmental organizations. However, these efforts are often in reaction to severe debris impacts, such as damage to sensitive coral ecosystems in the NWHI, or because of citizen awareness and support. Unlike some other marine pollution problems such as oil spills, there is no organized effort to assign roles and responsibilities for derelict gear recovery or cleanup. Under a "polluter pays" principle, fishery participants and the associated fishery management community should collectively take responsibility for the full spectrum of their impacts on the environment, including the fate of lost gear. Currently, however, any responsibility toward fishing gear seems to disappear the moment it becomes derelict.

A key difference between the loss of gear and most other marine pollution incidents is that "accidental loss" of fishing gear may be excusable and therefore no explicit infraction has occurred. In that sense, there is no fault and no individual "polluter" to pay for derelict gear recovery. However, the absence of a prosecutable infraction (except in extreme cases) could allow the entire community, rather than a culpable individual, to take responsibility and work toward solutions. Under this "no fault" approach, gear recovery programs that fit specific fishery and environmental circumstances can be developed. Box 4.5 describes a successful case of a gear retrieval program in Puget Sound that engages the responsible parties, both the fishermen and managers, in a "no fault" program. The Northwest Straits Commission's program could serve as a useful model for other fisheries communities facing gear removal.

Finding: Fishing is inherently hazardous and, of a necessity, entails some risk of gear loss despite all reasonable precautions. Because it is difficult for enforcement agencies to clearly differentiate between willful, preventable, and unpreventable gear losses, enforcement of a strict liability for gear losses would be problematic and could lead fishermen to underreport losses or obscure the location of gear losses.

BOX 4.5
Recovery of Derelict Fishing Gear in Puget Sound

In the state of Washington, the Northwest Straits Commission organized a "no fault" reporting program to facilitate the location, removal, and return or disposal of lost Dungeness crab pots and salmon gillnets in Puget Sound and the Straits of Juan de Fuca (Northwest Straits Commission, 2008). With grant support from the National Oceanic and Atmospheric Administration, the state, and private sources, a publicly reviewed protocol for gear removal, return, and disposal was developed. Among other considerations, the protocol addresses the "no fault" policy, liability, safety, habitat damage issues associated with gear recovery, disposal methods and costs, data collection, and a detailed guide for the removal processes. The "no fault" aspect of this program is that fishermen and the general public are encouraged to report the location of derelict fishing gear without being held liable for recovery and disposal costs.

Regardless of whether an infraction has occurred, current regulations do not include accountability measures for gear loss and fishermen and fisheries management organizations have few incentives and several disincentives to take responsibility for the impacts and for cleanup.

Recommendation: Fishery management organizations, if they adopt gear loss reporting and other accountability measures, should adopt a "no fault" policy regarding the documentation and recovery of lost fishing gear. Under this policy, local fishermen, state officials, and the public should work together to develop cost-effective DFG removal and disposal programs. These programs could be subsidized through user fees; a tax or deposit on trap tags, permits, or gear; public and private grants; or mitigation banking. Fishermen participating in removal efforts could receive financial credit, or at least be exempted from landfill tipping fees.

Adequate Reception Facilities for Fishing Gear

There are unique challenges to the proper disposal of fishing gear—both nonoperational waste gear and recovered DFG—associated with the lack of reception facilities for this gear once it is brought to shore. Unlike other waste streams, fishing gear is often very bulky and may require special handling and a lot of landfill space. Conversely, fishing ports are often small operations in remote areas, which can lead to high costs and limited options for solid waste management. When fishermen are required to pay full disposal costs for used fishing gear, used gear tends to accumulate in storage yards and they are discouraged from retrieving DFG. In some remote fishing ports, including Unalaska (Dutch Harbor), Alaska—the highest volume fishing port in the United States—landfill capacity is so constraining that disposal of trawl web costs $106 per ton if cut and bound into 1 cubic yd bundles and $106 per ton plus $500 per cubic yd if not cut and bound (City of Unalaska, 2008). While landfilling is often the least expensive and most feasible option for fishing gear disposal, there are several alternative technologies that are currently being pursued or have promise for future operations, including recycling, combustion with energy recovery, gasification, pyrolosis, and plasma arc systems (see Appendix E). There are a few ports that have already adopted alternative waste management strategies for fishing gear. For example, the Port of Honolulu maintains bins for free disposal of used fishing gear, including recovered derelict gear. Schnitzer Steel Hawaii provides hauling and pre-processing of the used gear, the state of Hawaii waives disposal fees, and Honolulu Power combusts the preprocessed wastes in an energy recovery facility (Rene Mansho, personal communication; Rodney Smith, personal communication; Howard Wiig, personal communication).

The feasibility of alternative fishing gear waste management options will depend on such conditions as reliability of the fishing gear and other waste streams, waste transportation costs, and difficulty in siting and permitting new facilities. It was beyond the scope of this study to analyze the potential of each of these technologies for various ports, but a summary of these various management options is provided in Appendix E.

Finding: Remote ports often have difficulty providing adequate port reception facilities for used fishing gear and recovered DFG.

Recommendation: The actual ability to receive used fishing gear and DFG should be incorporated into minimum standards in the assessment criteria for USCG certificates of adequacy for port reception facilities.

Recommendation: The Environmental Protection Agency, NOAA, and the U.S. Army Corps of Engineers, in cooperation with the fishing industry, ports, and fishery managers, should help fishing communities explore alternative strategies and technologies for management, disposal, and recycling of used and recovered DFG.

Finding: Disposal costs discourage proper disposal of used fishing gear and can also be a disincentive to DFG retrieval.

Recommendation: The Interagency Marine Debris Coordinating Committee and the NOAA Marine Debris Program should consider expanding the marine debris cleanup grants program to help offset the disposal costs for recovered DFG. Consideration should be given to dropping the 50 percent match requirement for DFG recovery and disposal programs, particularly for small remote communities.

Legal Recovery of Derelict Gear

There is a great deal of interest and ongoing effort by government agencies, fishing industry groups, conservation organizations, and others to retrieve and remove derelict gear from the environment. While in-the-water removal of DFG can be dangerous and requires significant attention to safety, other legal and financial challenges can pose even greater challenges and, in some cases, significantly hamper the ability to remove this gear.

First, there are often legal restrictions that prevent some vessels, including fishing vessels that would like to voluntarily participate in gear retrieval programs, from carrying DFG. For example, some nations, including the United States, have enacted laws (e.g., 46 U.S.C. § 2101(13), 46 U.S.C. § 3301(1)) which could prevent the retrieval of DFG under a compensatory scheme (even arguably payment of extra fuel costs incurred in the retrieval and transport to shore operation) as such an action could legally change the character of the vessel to one carrying "freight for hire," which then triggers other legal provisions including national inspection requirements and possibly cabotage[3] laws. Similarly, the New England Fishery Management Council noted that fishing vessel operators could be in violation of their fishing permits if the derelict gear that they recover and transport is not a gear type for which they hold a license endorsement. In practice, fishing vessels that have recovered such gear have been able to call USCG for authorization to transport it to port for disposal. Also, the North Pacific Fishery Management Council noted that removing DFG from habitat used by listed resources, such as Steller sea lions, could require a "take" permit.

Second, in many areas, static gear (particularly pots and traps) has an ownership component that enjoins tampering with or handling gear that belongs to others, even if that gear appears to be lost or abandoned. These prohibitions are intended to prevent theft of the gear or theft of its contents. For example, in Florida a fishing trap is considered personal property and cannot be removed by anyone except the owner or a licensed enforcement officer. In many states, development of a derelict trap removal program would require revision of statutes and regulations—a process that is often difficult and lengthy. In some states, this problem has been resolved by

---

[3]Cabotage refers to trade or transport within coastal waters.

declaring that traps left in the water during closed seasons are trash that can be salvaged or removed by fishermen, government officials, or the general public.

An additional impediment to the recovery of DFG is that, as soon as it has been removed from the environment, it becomes the possession of whoever removed it. The possessor is then liable for proper disposal. Because very few ports have free disposal services for used fishing gear or for recovered DFG, individuals and organizations that recover or transport lost or abandoned fishing gear may be liable for disposal costs which could be substantial (see previous discussion).

Finding: Some legal frameworks discourage or prevent the retrieval of DFG. In the United States, recovery of DFG may be inhibited by prohibitions against tampering with abandoned gear, the application of cabotage laws and burdensome certification requirements for vessels that transport DFG, and fishery regulations that prohibit vessels from carrying gear that is not a gear type permitted under their license endorsement.

Recommendation: USCG should work with other federal agencies, state officials, fishermen, and the public to revise regulations that inhibit the removal of DFG.

FISH AGGREGATING DEVICES

The growing use of a specific type of fishing gear—FADs—in pelagic purse seine fisheries raises questions about its potential impacts on both target and nontarget populations, as well as its potential to become marine debris. FADs, their use in fishing operations, and their potential impacts are defined and described below, focusing on FADs as marine debris. The legal status of and management options for FADs are also discussed.

What Are Fish Aggregating Devices?

For thousands of years, fishermen have exploited the tendency of fish to school beneath floating objects. These objects can be natural floating flotsam, such as logs and branches, dead marine organisms, and aquatic vegetation. They can also be man-made FADs, constructed from scrap lumber, rope, and discarded fishing gear; therefore, the cost of construction and the overall value is minimal. Man-made FADs are used in shallow coastal waters (depth 50–200 m) by artisanal fishermen to catch small pelagic fish (e.g., Philippines, Malaysia, Indonesia) and are also used heavily in offshore industrial purse seine fleets to catch large pelagic fish, mainly tuna (Box 4.6 describes the use of drifting FADs by the U.S. purse seine fleet). FADs may be anchored to the seafloor or allowed to drift (Food and Agriculture Organization of the United Nations, 2008c). This discussion will focus on man-made drifting FADs as they are typically constructed of bamboo with panels of synthetic webbing, plastic floats, synthetic ropes, and often include plastic sheeting. Moreover, man-made drifting FADs have the greatest potential to be lost or abandoned and become marine debris. Figure 4.1 illustrates the design of a typical drifting FAD.

Drifting FADs are usually deployed at or near the surface, but some are designed to fish in midwater with a small marker buoy on the surface. A growing number of drifting FADs are equipped with autonomous sonar buoys that can report GPS position, current speed, sea surface temperature, and sonar images of associated fish to catcher vessels, which

---

BOX 4.6
Use of Drifting Fish Aggregating Devices by
U.S. Fishing Vessels in the Western Pacific

The United States licenses a number of U.S. flag purse seine fishing vessels that use drifting fish aggregating devices (FADs) in the western Pacific to catch tuna. This fleet moved to the western Pacific from its traditional fishing grounds in the eastern tropical Pacific in the 1980s (Gillett et al., 2002). These vessels home-port in American Samoa, a U.S. territory, and many deliver their catches to tuna canneries located there. The U.S. purse seine vessels fish in the U.S. Exclusive Economic Zones (EEZs) of 16 Pacific island nations under the South Pacific Tuna Treaty between the Pacific Islands Forum Fisheries Agency and the United States. The fleet also operates within the U.S. EEZ around U.S. territories and island possessions in the western Pacific. In recent years, approximately 12–15 new super seiners have joined the fleet, under a joint venture with Taiwan and a waiver of the Jones Act (46 U.S.C. App. § 688) (David Itano, personal communication). In 2008, the National Oceanic and Atmospheric Administration published a notice of a possible control date after which new entrants to the purse seine fishery are not guaranteed a fishing license (73 Fed. Reg. 16619–16620 [March 28, 2008]). These vessels are subject to regulation under the Magnuson-Stevens Fishery Conservation and Management Act, the South Pacific Tuna Act (16 U.S.C. § 973 et seq.), the Western and Central Pacific Fisheries Convention Implementation Act (P.L. 109-479), and the High Seas Fishing Compliance Act (16 U.S.C. § 5501 et seq.), as well as to the licensing authority of the Pacific Islands Forum Fisheries Agency. U.S. regulations for this fishery are codified at 50 C.F.R. Part 300. These regulations define deploying and recovering FADs or associated electronic equipment as "fishing" and require vessel and gear identification marking. They do not currently require vessels to have a FAD management plan as a condition of their licenses. Issuance of these licenses, however, is subject to review and consultation under the Endangered Species Act (*Turtle Island Restoration Network v. NMFS*, 340 F.3d 969 [9th Cir. 2003]).

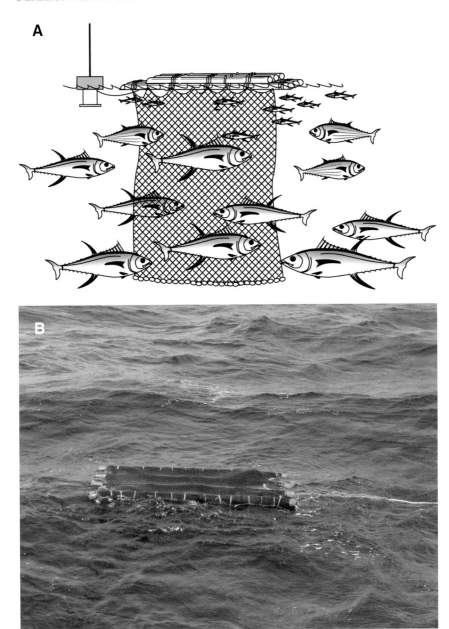

FIGURE 4.1 (a) Graphical representation of a typical drifting FAD (Itano et al., 2004; reprinted with permission from the Oceanic Fisheries Programme). (b) View of a drifting FAD from the surface (reprinted with permission from David Itano).

may be located thousands of miles away, to determine when FADs are ready for fishing (Itano, 2007a, b). Deep-water tuna vessels will set many drifting FADs at once, generally at 5- to 10-nautical-mile intervals. These FADs can be moved from one location to another in order to capitalize on the seasonal movement of target species. FADs can also be removed from the water during closed seasons, when the fish are not in the area, or during periods of bad weather. Because it may take two to five weeks for a new FAD to attract large fish, fishermen will "seed" FADs upcurrent of productive waters and leave them to "season" between trips. Once a FAD is seeded, it can be fished every day for several days or every 10–30 days, depending on the productivity of the waters (Itano, 2007a, b; Food and Agriculture Organization of the United Nations, 2008c; Martín Hall, personal communication). Sets on FADs begin approximately two hours before first light. A small auxiliary craft ties up to the FAD to slowly maneuver the object during the set, assess school density and depth, and deploy underwater bait attraction lights (if used) (Itano, 2007a, b). The purse seine vessel uses information from sonar, Doppler current meter, and sea state to position the vessel in the optimal orientation to the floating object to begin the set and slowly encircles the school of tuna with a purse seine (Itano, 2007a, b). After the set, the FAD may be removed from the water for maintenance or reseeding in another location. While FADs are valued and are claimed by specific vessels, ownership can be difficult to assess, and FADs are often lost through vandalism, theft, drifting beyond the preferred fishing area, and storms (Itano, 2007a; Martín Hall, personal communication; Dick Stevenson, personal communication).

Fish Aggregating Devices as Marine Debris

The above description of how FADs are used illustrates some of the ambiguities that arise in considering their transformation into DFG. FADs come in many different constructions, mostly of old fishing gear and waste material. It may be unclear to outside observers which fishing vessel owns which particular FAD; vessels may set on seasoned FADs seeded by others, and FADs can be expropriated by switching radio and satellite buoys. In addition, several vessels working together, whether through formal or informal agreements, may share radio- or satellite-buoy codes, thereby sharing FADs (Dick Stephenson, personal communication). Regardless, as argued in Box 4.7, FADs are a plastic-containing fishing gear and failure of a deploying vessel or cooperative fleet to retrieve a FAD it has set into the ocean, barring exception for accidental loss if all reasonable precautions were taken, is an intentional disposal of synthetic fishing gear and is in violation of MARPOL Annex V and implementing regulations.

BOX 4.7
When Are Fish Aggregating Devices Marine Debris?

Section 1.7.11 of the *Guidelines for the Implementation of Annex V of MARPOL* defines "fishing gear" as "any physical device or part thereof or combination of items that may be placed on or in the water with the intended purpose of capturing, or controlling for subsequent capture, living marine or freshwater organisms" (International Maritime Organization, 2006b). Fish aggregating devices (FADs) meet this definition of fishing gear. Under MARPOL Annex V, unless they are composed entirely of natural materials, FADs also fall under the category of synthetic fishing gear. "Wastes" is defined in the *Guidelines* as "useless, unneeded, or superfluous matter which is to be discarded" (International Maritime Organization, 2006b). When a fishing captain makes a decision not to retrieve a FAD or cuts the radio or satellite beacon from it, the FAD has become waste fishing gear; leaving it in the sea at that point constitutes disposal. Some might argue that because the FAD is still aggregating, it is still fishing and is not waste and has not been disposed of. But fishing vessels do not leave fishing gear in the water to continue to fish for other unspecified vessels. The better interpretation is that, when the FAD is no longer aggregating fish on behalf of the vessel that deployed it (or other vessels that are part of its company or fishing association) and the captain decides not to retrieve the FAD, it is waste fishing gear that has been intentionally, not accidentally, abandoned in the sea, thus constituting a disposal.

Finding: There has been confusion over the legal status of FADs in relation to marine debris. However, under MARPOL and Annex V definitions, FADs become DFG when the captain of the vessel that last deployed the FAD decides not to retrieve it. This constitutes an illegal disposal under MARPOL Annex V and APPS if the FAD includes synthetic ropes, webbing, or other plastics. Transfers of FADs to other vessels, by agreement or appropriation, complicates attributing the discharge to a particular vessel.

Recommendation: NOAA should modify the federal regulations for U.S. tuna purse seine vessels to clarify the circumstances under which FADs become illegal discharges. Within international legal frameworks, the United States should encourage IMO and RFMOs to provide similarly explicit definitions of "accidental losses" and "reasonable precautions" to clarify the circumstances under which FADs constitute illegal discharges of marine debris.

Recommendation: RFMOs should devise regulations to exert greater control on the use, deployment, and retrieval of FADs to reduce the potential for FADs to become DFG. RFMOs should hold fishing fleets, nations, or the collection of all RFMO-licensed vessels responsible for retrieving all deployed FADs and should apply accountability measures such as loss of fishing privileges in RFMO waters. In turn, nations could potentially require retrieval of FADs by the vessel or fleet. In the United States, USCG should amend regulations implementing APPS to meet the intent of MARPOL Annex V and ensure that vessels fishing within U.S. waters and U.S. vessels fishing anywhere are held accountable to these standards.

Regional Fisheries Management Organizations

Within the past decade, there has been an increasing concern that derelict or lost FADs are contributing to the marine debris problem and some evidence exists to support this claim (Donohue, 2005). The ability to infer the extent to which derelict FADs are contributing to the marine debris problem is hampered by a lack of information on FAD use and their contribution as components of the DFG stream. It is clear that many more FADs are deployed each year than are retrieved by vessels and, therefore, at-sea circulating FAD numbers may be increasing (Martín Hall, personal communication; Dick Stephenson, personal communication). Consequently, fishery resource managers and scientists recognize that considerable data on FAD use and retrieval are needed, not only to better understand their role as sources of DFG but also to understand their impact on managed fisheries.

To evaluate the contribution of FADs to the marine debris problem, the committee sought information about the number of FADs deployed and lost, thereby assessing the possible source stream of derelict FADs. Evidence shows that the use of FADs has significantly increased in pelagic fisheries and FADs are now widely distributed in tropical and subtropical waters globally, contributing to more than half of the worldwide tuna catch (Hallier and Gaertner, 2008). The use of FADs is particularly prevalent in the Indian Ocean (Itano, 2007b). It is estimated that there may be tens of thousands of FADs deployed throughout the oceans of the world (Dick Stephenson, personal communication).

Similarly, an assessment of existing regulations is helpful in understanding concerns about FADs that have led to FAD management. As most FAD fisheries occur on the high seas, it is reasonable to expect management of FADs to take place within RFMOs; however, there are currently very few international controls on FADs. In most international fisheries, FADs are deployed without any regulations on the number deployed, where they are deployed, identification markings, reporting of how often they are set on, whether they are retrieved, or reporting of the number lost and the circumstances of their loss or abandonment. While there is still little documentation on FAD use and minimal regulation, the available information is summarized below for the four RFMOs known to have large-scale FAD use.

Inter-American Tropical Tuna Commission

The IATTC database provides the most detailed information on FAD fishing of any regional fishery management organization. The tuna purse seine fleet fishing in the eastern tropical Pacific (ETP), managed by IATTC, has grown from 125 vessels in 1961 to 225 in 2006; the majority of the vessels are large (greater than 363 metric tons capacity) (Inter-American Tropical Tuna Commission, 2007a). Fishing on FADs or floating objects is one of three ways to capture tuna in ETP (the others involve encircle-ment of free-swimming schools and schools associated with dolphins) (Figure 4.2a). While natural flotsam (i.e., things *found* floating as opposed to things *deployed* with a purpose) is still opportunistically used as a FAD when encountered, ETP fishermen almost exclusively fish on man-made drifting FADs. FADs have been widely used in the ETP purse seine fishery for almost 15 years, and their relative importance has increased during this period, while that of flotsam has decreased, as shown by the data in Figure 4.2b.

In the ETP tuna purse seine fishery, vessels generally deploy 50–75, but in some cases up to 330, FADs annually (Altamirano et al., 2004; Martín Hall, personal communication); in 2006, the entire IATTC fleet deployed 8,188 FADs, and 8,721 FADs were deployed in 2007 (Martín Hall, personal communication). Figure 4.2 shows that the number of sets on FADs has been increasing since 1992, suggesting an increasing reliance on FADs in the fishery.

ETP FADs are generally equipped with a satellite buoy or radio beacons. For the most part, as long as the satellite buoy is still functional and the FAD can be relocated and retrieved at a profit, fishermen will return to a FAD both to fish and to retrieve equipment (Dick Stephenson, personal communication). The extent to which FADs are removed from the water during fishing closures or the end of the fishing season is undocumented. Information on the number of FADs retrieved by ETP vessels each year is very limited. For 2006, 6,163 FADs were retrieved out of 8,188 deployed, and for 2007, 7,769 FADs were retrieved out of 8,721 deployed (Martín Hall, personal communication). Of the 2,025 and 925 deficit for 2006 and 2007, respectively, many FADs are probably still deployed fishing, while others may have been "appropriated" by other vessels and are still being used; the remainder have been abandoned and, to the extent that they were constructed of synthetic rope webbing and plastics, the failure to retrieve them constitutes a discharge in viola-tion of MARPOL Annex V. Anecdotal evidence, provided by an experi-

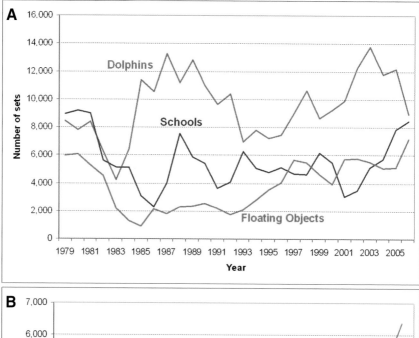

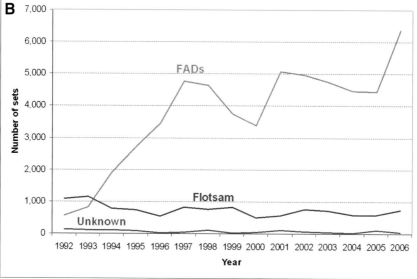

FIGURE 4.2 (a) Estimated number of sets on FADs, by type, made by Class 6 purse seine vessels (capacity greater than 343 metric tons) in the eastern Pacific Ocean (reprinted with permission from the Inter-American Tropical Tuna Commission). (b) Estimated number of sets on floating objects, by type of object, encountered by the purse seine fleet in the eastern Pacific Ocean. Flotsam are objects *found* floating whereas FADs are intentionally deployed objects (reprinted with permission from the Inter-American Tropical Tuna Commission).

enced captain, suggests that the appropriation rate—either removal of the entire FAD or the satellite or radio beacon—may be substantial (Dick Stephenson, personal communication).

Researchers at IATTC have proposed to give each FAD a unique code that could be recorded by observers and could be very useful for a variety of scientific purposes, ranging from the stock assessment of target and nontarget species to the drift of the FADs. While IATTC does not have any specific regulation on FAD usage, the Antigua Convention of IATTC (Article VII, 1(g) and (k)) contains language specific to reducing bycatch and developing environmentally safe fishing gear, which could be applicable to FADs:

(g) adopt appropriate measures to avoid; reduce; and minimize waste, discards, catch by lost or discarded gear, catch of non-target species (both fish and non-fish species), and impacts on associated or dependent species, in particular endangered species; . . .

(k) promote, to the extent practicable, the development and use of selective, environmentally safe and cost-effective fishing gear and techniques and such other related activities, including activities connected with, inter alia, transfer of technology and training; . . .

(Inter-American Tropical Tuna Commission, 2003).

Western and Central Pacific Fisheries Commission

The tuna fishery in the western and central Pacific Ocean, managed by WCPFC, is diverse, ranging from small-scale artisanal operations in the coastal waters of Pacific states to large-scale industrial operations in both the EEZs of Pacific states and on the high seas (Figure 4.3). Over the past five years, the trend in total tuna caught has been increasing, primarily due to increases in purse seine fishery catches. During 2006, the purse seine fishery in the western and central Pacific Ocean accounted for an estimated 1.5 million metric tons (72 percent of the total catch—only 12,000 metric tons less than the record catch of 2005) (Williams and Reid, 2006).

There are about 225 purse seine vessels fishing in the western and central Pacific Ocean; however, this estimate does not include Indonesian and Filipino domestic purse seine/ringnet fleets which together account for over 1,000 vessels (Williams and Reid, 2006). Sets on floating objects (logs and FADs) accounted for about 51 percent of all reported WCPFC sets during 2006 (Williams and Reid, 2006; David Itano, personal communication). Of the associated set types, log sets have been favored over drifting FAD sets by most purse seine fleets in recent years, with the exception being the U.S. fleet, which continues to operate in more eastern (and southern) areas of the western and central Pacific Ocean concentrat-

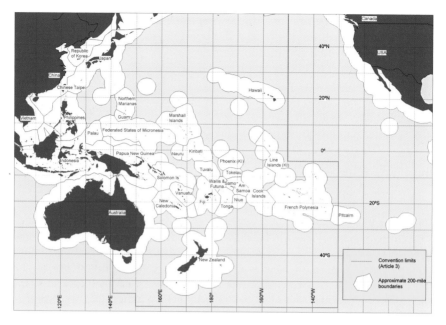

FIGURE 4.3 The western and central Pacific Ocean, the eastern Pacific Ocean, and the Western and Central Pacific Fisheries Commission Convention Area (reprinted with permission from the Western and Central Pacific Fisheries Commission).

ing on drifting FAD sets (69 percent in 2006 according to available log-sheet data) (Williams and Reid, 2006). Overall, information on how many FADs are deployed and the rate of FAD loss, appropriation, and recovery is unknown for the WCPFC fleet.

While WCPFC does not have any regulations specific to the use of FADs, the Convention on the Conservation and Management of Highly Migratory Fish Stocks in the Western and Central Pacific Ocean contains language that specifically requires measures to minimize "catch by lost or abandoned gear" and could also be applied to derelict FADs:

> adopt measures to minimize waste; discards; catch by lost or abandoned gear; pollution originating from fishing vessels; catch of non-target species, both fish and non-fish species;...and impacts on associated or dependent species, in particular endangered species and promote the development and use of selective, environmentally safe, and cost-effective fishing gear and techniques (Article 5(e)).

International Commission for the Conservation of Atlantic Tunas

A study published by Ménard et al. (2000) estimates that the total number of FADs with radio or satellite buoys used by the 45 purse seiners landing in Abidjan (Côte d'Ivoire) in 1998 might exceed 3,000. The FAD seeding area ranges from 0 to 20°W and generally does not exceed 2°S as a southern limit, corresponding to the westward South Equatorial Current (Ménard et al., 2000). Within the International Commission for the Conservation of Atlantic Tunas (ICCAT), information is completely lacking on the number of FADs deployed, the number of sets on any given FAD, and the number of FADs retrieved, lost, or appropriated each year.

Within ICCAT, control of FADs rests with two provisions. First, ICCAT requires that all fishing vessels and fishing gear have identifiable markings in accordance with generally accepted standards (International Commission for the Conservation of Atlantic Tunas, 2003). The second is a moratorium on FAD fishing in given areas, which was intended to reduce fishing mortality on bigeye tuna, particularly juvenile bigeye, but may have a collateral benefit in reducing the number of FADs (and therefore the number that could become debris). The "Agreement of the Community Producers of Frozen Tuna for the Protection of Tunas in the Atlantic Ocean" established a voluntary regulation prohibiting anchoring or fishing under floating objects in a wide area of the Atlantic Ocean, between the African coast and 20°W and 5°N and 4°S, from November 1997 to January 1998. The agreement was continued during the same months of 1998 and 1999 (International Commission for the Conservation of Atlantic Tunas, 2001). In 2004, the Commission adopted a substitute time–area closure, which entered into force in mid-2005 (International Commission for the Conservation of Atlantic Tunas, 2004). This measure closes fishing by purse seiners and bait boats during the month of November inside the "Piccolo" area, a small subregion (less than 25 percent) of the original moratorium area. The Piccolo area is defined as 10°–20°W and 0°–5°N (Figure 4.4).

Indian Ocean Tuna Commission

Since the 1990s, FAD usage by European Union purse seine fleets has increased significantly in the Indian Ocean (Morón et al., 2001), particularly in the Somalia gyre and around the Seychelles plateau, where FADs are the dominant fishing mode (Itano, 2007b).

Here, drifting FADs lack surface rafts or floatation, aside from some purse seine corks and the radio or satellite buoy, and are instead carefully ballasted plastic oil drums suspended below the surface with nylon netting hanging beneath the drums. This style of FAD is popular as it reduces

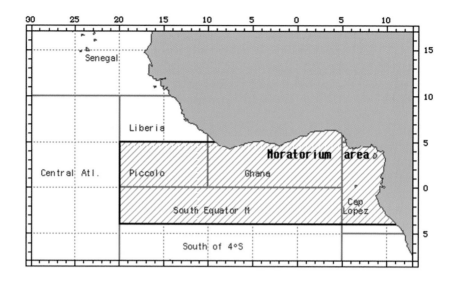

FIGURE 4.4 Area of the current FAD moratorium (hatched area) and the proposed time–area closure (i.e., "Piccolo") (International Commission for the Conservation of Atlantic Tunas, 2001; reprinted with permission from the International Commission for the Conservation of Atlantic Tunas).

the surface visibility of the FAD and therefore its rate of appropriation by other vessels.

The Spanish purse seine fleet operating in the western Indian Ocean is assisted by supply (or tender) vessels; these vessels, in addition to other duties, may search for FADs and logs, build or repair FADs, assess tuna abundance on other floating objects it encounters, and appropriate productive FADs belonging to other vessels (Arrizabalaga et al., 2001). Tender vessels clearly improve the ability of fishing associations to utilize FADs. Consequently, the added efficiency has led to the banning of their use in the Pacific and the Atlantic tuna fisheries; therefore, the Indian Ocean Tuna Commission (IOTC) is the only fleet with tender vessels that service FADs (Itano, 2007a).

Skipper surveys from French and Spanish purse seine vessels operating in the western Indian Ocean estimated the total number of actively monitored FADs at approximately 2,100 at any given time (Moreno et al., 2007). IOTC views this number as a highly dynamic estimate, as FADs can sink or be appropriated by other purse seiners and have a lifetime between a few days to several months. In order for IOTC to better understand the fishing effort within the Indian Ocean, more information is needed on the activities of supply vessels and the use of FADs. Therefore, IOTC is now requesting that its members provide

- the number and characteristics of supply vessels operating under or assisting purse seine vessels operating under each nation's flag, or licensed to operate in a nation's exclusive economic zones;
- the level of activity of supply vessels, including number of days at sea by 1° grid area and on a monthly basis; and
- data on the total number and type of FADs operated by a nation's fleet by 5° grid area and on a monthly basis (Indian Ocean Tuna Commission, 2007).

Despite this requirement, within the IOTC fleet, information is completely lacking on the number of FADs deployed or carried by each vessel; the number of sets on any given FAD; and the number of FADs retrieved, appropriated, or lost each year.

Given the information collected on FAD use in the Pacific, Atlantic, and Indian Oceans, it is clear that FADs could contribute a substantial amount of marine debris. However, much more information is needed to fully understand the extent of this problem.

Other Impacts

While the committee's charge was to evaluate the role of drifting FADs in the generation of marine debris, the concern over FADs is primarily focused on their ecological impact, both on target fisheries species and on pelagic species overall. These broader concerns do not go away after FADs have been lost or otherwise abandoned—FADs as DFG can be expected to exercise an ecological impact on target and nontarget species and on benthic and littoral ecosystems when they sink or wash ashore. Therefore, it is useful to briefly discuss these other impacts of FADs.

The widespread use of FADs has shifted the pattern of fishery exploitation of tunas over the past 20 years. In the Atlantic and Indian Oceans, approximately 75 percent of skipjack tuna (*Katsuwonus pelamis*), 35 percent of yellowfin tuna (*Thunnus albacares*), and 85 percent of bigeye tuna (*T. obesus*) catches reported by purse seine fisheries are made in the vicinity of FADs (Fonteneau et al., 2000). In all oceans, the majority of yellowfin and bigeye tuna caught in association with FADs are juveniles. Therefore, fishing on FADs may alter the age structure of some pelagic tuna populations by removing juveniles over mature adults (Gates and Gysel, 1978; Fonteneau et al., 2000; Schlaepfer et al., 2002; Hallier and Gaertner, 2008).

Some scientists are concerned that FADs may function as an eco-logical trap: a situation where population growth is reduced as a result of individuals choosing a maladaptive habitat (Gates and Gysel, 1978; Schlaepfer et al., 2002). It is hypothesized that this situation could arise if individuals are misled by environmental cues that lead them to settle in habitats that are substandard for reproduction and survival (Battin, 2004; Robertson and Hutto, 2006). Association with FADs may alter the natural movements of fractions of tuna stocks and thereby artificially increase the natural mortality rate or reduce the intrinsic growth rate, reducing the productivity of tuna populations (Hallier and Gaertner, 2008). For example, studies in the Atlantic and Indian Oceans indicate that tuna associated with drifting FADs were less healthy, have slower growth rates, and are in poorer condition than those in free schools (Hallier and Gaertner, 2008). Also, tuna associated with FADs have significant changes in migratory direction and displacement rates relative to tuna in free schools (Hallier and Gaertner, 2008).

Studies in recent years, especially within the ETP tuna fishery, indi-cate that FAD fishing bycatch (i.e., discards of small tuna and nontarget species) can be up to 50 percent of the total catch (Inter-American Tropical Tuna Commission, 2007b). One study reported that almost 20 percent of the tuna caught under FADs are discarded because they are below the market minimum requirement for size or condition (Inter-American Tropical Tuna Commission, 2007b). Bycatch of small tuna and other spe-cies contributes to discarded, unreported, or underreported catch and may represent a significant source of undocumented fishing mortality. In addition to undersized tuna, FAD-associated bycatch includes large pelagic fishes (e.g., mahi-mahi, rainbow runner, yellowtail) and under-sized billfishes (Fam. Istiophoridae), anchovies (Fam. Engraulidae), her-rings and sardines (Fam. Clupeidae), and grunts (Fam. Haemulidae). Entanglement of sea turtles in drifting FADs has been noted as an area of special concern by scientists and the purse seine industry (Delgado de Molina et al., 2006). Likewise, the bycatch of several species of sharks in association with FADs is an increasing concern due to declines in their populations (Hall, 1994).

Improving the Understanding and Management of Fish Aggregating Devices

To date, very little is known about the total number of FADs in the world's oceans, the number of vessels that fish on or use FADs, the number of FADs deployed by fishing vessels, whether and with what frequency FADs are recovered, the frequency with which individual FADs are set upon, the total number of sets on FADs, and the expropriation and loss rate of FADs. RFMOs have a role in collecting and a need for improved data on FADs to achieve their goals of sustainable international fisheries with minimal environmental impact.

Currently, at-sea observer programs are the best means to collect specific data on FADs and their use. However, tracking and identifying drifting FADs can be difficult, with FADs taken onboard, modified, and in some cases appropriated from other vessels and provided with a different radio or satellite buoy. Greater control and documentation of FADs is needed if FAD deployment, usage, and loss are ever to be understood.

The IATTC observer program has the most complete record of FAD use through its Flotsam Information Record (FIR) program (Figure 4.5). FIR contains the key points to consider when describing and tracking floating objects. The form includes parameters such as time and location, description and dimension of the FAD and its components (including vertical appendages and associated electronics), how the FAD was located, and information on the origin or ownership of the FAD. FIR also describes whether the FAD is left in the water and any significant alterations or enhancements that may have been made. The FIR program could serve as a model system for collecting data on FADs for other regional fishery management organizations.

Similarly, RFMOs have a role in improving regulations and management of FADs. In December 2007, WCPFC considered, but has yet to implement, the most comprehensive resolution on FAD use. The resolution would prohibit FAD fishing between either July through September or October through December in the EEZs and the areas beyond national jurisdiction within the area bounded by 20°N and 20°S. The resolution has an exemption for purse seiners home ported in the Philippines and operating on the high seas off the coast of the Philippines, which are entirely dependent on FAD sets, but requires the Philippines to implement its national tuna plan, which limits the number of FADs to 25 FADs per purse seine vessel and to provide the national tuna plan for review and endorsement in 2008 by WCPFC.

Even more notable was the requirement that parties submit to WCPFC management plans for the use of FADs within their jurisdictional waters and by their vessels on the high seas containing the following elements:

- limits on the number of licensed FADs;
- design, operation, and maintenance of FADs;
- application process for deployment of FADs;
- location of FADs and reporting;
- marking of FADs;
- location in relation to navigational routes and shipping;
- closed areas;

Inter-American Tropical Tuna Commission
FLOTSAM INFORMATION RECORD (FIR)

Trip Number	Object No.	Count No.	Set No.	YY	DATE MM	DD	TIME	LATITUDE	N/S	LONGITUDE	W

A. COMPONENTS (check all that are applicable)

	As found		As left
Tree	[ ]	1	[ ]
Dead animal _____	[ ]	2	[ ]
Chain / cable / rings / weights	[ ]	3	[ ]
Cane / bamboo	[ ]	4	[ ]
Bait container / bait	[ ]	5	[ ]
Cord / rope	[ ]	6	[ ]
Floats / corks	[ ]	7	[ ]
Artificial light for attracting fish	[ ]	8	[ ]
Netting material	[ ]	9	[ ]
Sacks / bags	[ ]	10	[ ]
Planks / pallets / plywood / spools	[ ]	11	[ ]
Metal drum / plastic drum	[ ]	12	[ ]
PVC or other plastic tubes	[ ]	13	[ ]
Plastic sheeting	[ ]	14	[ ]
Unknown	[ ]	15	[ ]
Other _____	[ ]	16	[ ]

B. LOCATING EQUIPMENT (check all that are applicable)

	As found		As left
Flag	[ ]	1	[ ]
Satellite buoy	[ ]	2	[ ]
Buoy, corks, etc.	[ ]	3	[ ]
Lights	[ ]	4	[ ]
Radio transmitter / beeper	[ ]	5	[ ]
Radar reflector	[ ]	6	[ ]
Unknown	[ ]	7	[ ]
Other _____	[ ]	8	[ ]

C. LOCATING METHOD (check only ONE)

Radar	[ ]	1
Direction finder	[ ]	2
Satellite	[ ]	3	check
Visual – the object itself	[ ]	4	only
Visual – birds	[ ]	5	one
Not applicable	[ ]	6
Unknown	[ ]	7
Other _____	[ ]	8

D. IF THERE IS NETTING ON THE OBJECT:

	Yes	No	Unk
Netting hanging from the object?	[ ]	[ ]	[ ]

Estimated area of hanging netting (m²) [ ]

Predominant mesh size (inches) [.]

E. OTHER DATA

	Yes	No	NA	Unk
Bait container refilled?	[ ]	[ ]	[ ]	[ ]
Fauna entrapped? _____	[ ]	[ ]	[ ]	[ ]

Maximum depth of the object (m) [.]

Dimensions (m) [.] [.] [.]

Water clarity   Clear [ ]   Turbid [ ]   Very turbid [ ]

% epibiota [ ]   Tag number [ ]

F. CAPABILITY OF TRANSMITTING EQUIPMENT (check all that are applicable)

	As found		As left
Direction to the object	[ ]	1	[ ]
Geographic position of the object	[ ]	2	[ ]
Water temperature	[ ]	3	[ ]
Tuna quantity	[ ]	4	[ ]
Tuna species	[ ]	5	[ ]
Unknown	[ ]	6	[ ]
Other _____	[ ]	7	[ ]

G. PRIOR ORIGIN OF OBJECT (check only ONE)

Your vessel – this trip	[ ]	1
Your vessel – previous trip	[ ]	2
Deployed	[ ]	3
Other vessel – with owner consent	[ ]	4	check
Other vessel – no owner consent	[ ]	5	only
Drifting object found	[ ]	6	one
Unknown	[ ]	7
Other _____	[ ]	8

H. EXPERIMENTAL EQUIPMENT (continue on back)

IATTC FIR 08/2005

FIGURE 4.5 IATTC Flotsam Information Record (FIR) card (reprinted with permission from the Inter-American Tropical Tuna Commission).

I.a. OVERHEAD VIEW (Include dimensions)	I.b. SIDE VIEW (Include dimensions)

J. ADDITIONAL COMMENTS

IATTC FIR 08/2005

FIGURE 4.5 (Continued)

- deployment of FADs in archipelagic waters;
- effect of FAD fishing by purse seine vessels on tuna longline fishing;
- monitoring of the FAD fishery;
- effect of FAD fishing on sizes of tuna taken;
- effect of FAD fishing on bycatch species;
- reporting requirements for FAD fishing;
- reporting of species mix in FAD fishing;
- reporting of bycatch in FAD fishing;
- reporting of utilization of bycatch;
- conflict resolution in relation to FADs;
- license status of vessels in relation to areas of FAD deployment;
- replacing of lost FADs;
- access to FAD areas;
- confidentiality of FAD position information; and
- number of tender vessels per catcher vessel.

Implementation of this resolution would be an important step toward greater control and understanding of FADs. Information collected from FAD management plans could be used to more effectively evaluate the role of FADs in the generation of marine debris.

In 1999, IATTC considered (but failed to adopt) the following measures to reduce bycatch and adverse impacts of FADs on the tuna resource:

- limits on the depth of FADs;
- limits on the number of sets on FADs and floating objects;
- limits on the number of FADs that a vessel can carry;
- analysis of the effects of the use of bait with FADs;
- seasonal or area bans or closures on the use of FADs; and
- modification of the FAD design (Inter-American Tropical Tuna Commission, 1999).

These measures, if adopted by IATTC, could provide a framework for greater control over FAD fishing in ETP.

Finding: Currently, there is very little control or data on FADs in international fisheries.

Recommendation: The United States should take a leadership role by

- requiring that its own purse seine fleet submit a FAD management plan incorporating the plan elements proposed by WCPFC;

- encouraging RFMOs to adopt requirements for FAD management plans; and
- using port state jurisdiction in its territories to limit access to vessels flying the flag of countries that fail to require their vessels have a FAD management plan.

Recommendation: RFMOs should adopt measures and manage FADs in such a way that the ownership of those FADs is clear. RFMOs should

- control the number of FADs through chips, marking, tags, or other means to limit the number of FADs that can be carried and deployed by a vessel;
- acquire more information to characterize FAD usage in each of the agreement areas;
- adopt resolutions requiring parties to provide information on FAD use by vessel, including the number of sets on FADs, the number of FADs carried and deployed, and FAD retrieval, loss, and appropriation rates; and
- establish mechanisms to gather information on FADs including reports from parties, vessel logbooks, and observer programs. At a minimum, RFMOs need to collect and report annual data on the number of FADs deployed, the number returned to shore, the number lost, and an annual estimate of the number currently being fished.

Finding: Replacement of plastic components and synthetic ropes and webbing used to construct FADs with readily degradable materials such as natural fibers would lessen the adverse impacts of FADs that become marine debris.

Recommendation: RFMOs should support the development of FAD designs that do not incorporate persistent synthetic or scrap materials but instead include materials that will self-destruct, readily biodegrade, mitigate entanglement, and provide an incentive for FADs to be maintained and regularly retrieved. RFMOs should also prevent the use of synthetic and scrap material in FADs through regulation.

CONCLUSION

The following finding and recommendation express overarching concepts discussed in the previous findings and recommendations in Chapter 4.

Overarching Finding: DFG and abandoned or lost FADs fall under both MARPOL Annex V (and corresponding domestic laws) and fisheries management treaties and regulations. This overlap has complicated implementation of measures to prevent and reduce these sources of debris. Current regulations do not include accountability measures for gear loss, and fishermen and fisheries management organizations have few incentives and several disincentives to take responsibility for the impacts and for cleanup. Inadequate port facilities and high disposal costs are an impediment to disposal of waste and DFG.

Overarching Recommendation: MARPOL Annex V (and corresponding domestic law) and international and domestic fisheries treaties and regulations should be revised to clearly identify and prohibit preventable losses of fishing gear, including FADs. IMO, FMCs and fishery management organizations, and other relevant entities should incorporate gear accountability measures and facilitate proper disposal of fishing gear, including FADs.

References

Acheson, J.M. 1977. The lobster fiefs: Economic and ecological effects of territoriality in the Maine lobster industry. *Human Ecology* 3:183-207.

Adler, J. 1987. Our befouled beaches: Condoms, styrofoam, and germs litter the sea. *Newsweek*, 27 July, 50.

Altamirano, E., M.A. Hall, and N.W. Vogel. 2004. Sightings of discarded fishing gear in the Eastern Pacific Ocean. In *Proceedings of the Seminar Derelict Fishing Gear and Related Marine Debris: An Educational Outreach Seminar Among APEC Partners*. Asia–Pacific Economic Cooperation, Singapore.

American Association of Port Authorities. 2006. *Glossary of Maritime Terms*. [Online]. Available: http://www.aapa-ports.org/Industry/content.cfm?ItemNumber=1077&&navItemNumber=545 [August 29, 2008].

Andrady, A.L. 1990. Environmental degradation of plastics under land and marine exposure conditions. In *Proceedings of the Second International Conference on Marine Debris*, Shomura, R.S. and M.L. Godfrey (eds.). U.S. Department of Commerce, Washington, DC.

Appleyard, E. 2004. *CCAMLR's Approach to Marine Debris Management in the Antarctic*. [Online]. Available: http://www.wpcouncil.org/documents/APECSeminar/Panel%204-%20International%20and%20Domestic%20Regulatory%20Structures/Presentation%20by%20Mr.%20Eric%20Appleyard.pdf [August 4, 2008].

Arnould, J.P.Y. and J.P. Croxall. 1995. Trends in entanglement of Antarctic fur seals (*Arctocephalus gazella*) in man-made debris at South Georgia. *Marine Pollution Bulletin* 30:707-712.

Arrizabalaga, H., J. Ariz, X. Mina, A. Delgado de Molina, I. Artetxe, P. Pallarés, and A. Iriondo. 2001. *Analysis of the Activities of Purse Seine Supply Vessels in the Indian Ocean from Observers Data*. Indian Ocean Tuna Commission, Mahé, Seychelles.

Asoh, K., T. Yoshikawa, R. Kosaki, and E. Marschall. 2004. Damage to cauliflower coral by monofilament fishing lines in Hawaii. *Conservation Biology* 18(6):1645-1650.

Azzarello, M.Y. and E.S. Van Vleet. 1987. Marine birds and plastic pellets. *Marine Ecology Progress Series* 37:295-303.

141

Bader, H. 1998. *Who Has the Legal Right to Fish?: Constitutional and Common Law in Alaska Fisheries Management*. University of Alaska Sea Grant, Fairbanks.

Balazs, G.H. 1978. A hawksbill turtle in Kanehoe Bay, Oahu. *Elepiao* 38(11):128-129.

Ballance, A., P.G. Ryan, and J.K. Turpie. 2000. How much is a clean beach worth? The impact of litter on beach users in the Cape Peninsula, South Africa. *South African Journal of Science* 5:210-213.

Barnard, D.R. 2008. *Biodegradable Twine Report to the Alaska Board of Fisheries*. Alaska Department of Fish and Game, Anchorage, Alaska.

Barnes, D.K.A. 2002. Invasions by marine life on plastic debris. *Nature* 416:808-809.

Barnes, D.K.A. 2005. Drifting plastic and its consequences for sessile organism dispersal in the Atlantic Ocean. *Marine Biology* 146:815-825.

Battin, J. 2004. When good animals love bad habitats: Ecological traps and the conservation of animal populations. *Conservation Biology* 18:1482-1491.

Bavestrello, G., C. Cerrano, D. Zanzi, and R. Cattaneo-Vietti. 1997. Damage by fishing activities to the Gorgonian coral *Paramuricea clavata* in the Ligurian Sea. *Aquatic Conservation: Marine and Freshwater Ecosystems* 7:253-262.

Bishop, R.C. 1982. Option value: An exposition and extension. *Land Economics* 58:1-15.

Bograd, S.J., D.G. Foley, F.B. Schwing, C. Wilson, R.M. Laurs, J.J. Polovina, E.A. Howell, and R.E. Brainard. 2004. On the seasonal and interannual migrations of the transition zone chlorophyll front. *Geophysical Research Letters* 31(L17204):1-5.

Boland, R.C. and M.J. Donohue. 2003. Marine debris accumulation in the nearshore marine habitat of the endangered Hawaiian monk seal, *Monachus schauinslandi* 1999-2001. *Marine Pollution Bulletin* 46:1385-1394.

Boland, R., B. Zgliczynski, J. Asher, A. Hall, K. Hogrefe, and M. Timmers. 2006. Dynamics of debris densities and removal at Northwestern Hawaiian Islands coral reefs. *Atoll Research Bulletin* 543:461-470.

Bourne, W.R.P. 1976. Seabirds and pollution. In *Marine Pollution*, Johnston, R. (ed.). Academic Press, London.

Bourne, W.R.P. 1977. Nylon netting as a hazard to birds. *Marine Pollution Bulletin* 8(4):75-76.

Brainard, R.E., D.G. Foley, and M.J. Donohue. 2000. Origins, types, distribution and magnitude of derelict fishing gear. In *Proceedings of the International Marine Debris Conference on Derelict Fishing Gear and the Ocean Environment*, McIntosh, N., K. Simonds, M. Donohue, C. Brammer, S. Manson, and S. Carbajal (eds.). Hawaiian Islands Humpback Whale National Marine Sanctuary, U.S. Department of Commerce.

Breen, P.A. 1987. Mortality of Dungeness crabs caused by lost traps in the Fraser River estuary, British Columbia. *North American Journal of Fisheries Management* 7:429-435.

Breen, P.A. 1990. A review of ghost fishing by traps and gill nets. In *Proceedings of the Second International Conference on Marine Debris*, Shomura, R.S. and M.L. Godfrey (eds.). U.S. Department of Commerce, Washington, DC.

Brown, G., Jr., and J.H. Goldstein. 1984. A model for valuing endangered species. *Journal of Environmental Economics and Management* 11:303-309.

Brown, J. and G. Macfayden. 2007. Ghost fishing in European waters: Impacts and management responses. *Marine Policy* 31:488-504.

Bush Administration. 2004. *U.S. Ocean Action Plan: The Bush Administration's Response to the U.S. Commission on Ocean Policy*. [Online]. Available: http://ocean.ceq.gov/actionplan.pdf [August 27, 2008].

California Coastal Commission. 2008. *The Problem with Marine Debris*. [Online]. Available: http://www.coastal.ca.gov/publiced/marinedebris.html [July 9, 2008].

California Ocean Protection Council. 2007. *Resolution of the California Ocean Protection Council on Reducing and Preventing Marine Debris*. [Online]. Available: http://www.resources.ca.gov/copc/02-08-07_meeting/0702COPC05_MarineDebris_Resolution.pdf [July 16, 2008].

California Ocean Protection Council. 2008. *An Implementation Strategy for the California Ocean Protection Council Resolution to Reduce and Prevent Ocean Litter.* [Online]. Available: http://www.resources.ca.gov/copc/docs/FINAL_DRAFT_IMPLEMENTATION_ STRATEGY_compressed.pdf [September 11, 2008].

California Regional Water Quality Control Board. 2001. *Attachment A: Amendment to the Water Quality Control Plan for the Los Angeles Region to Incorporate a Total Maximum Daily Load for Trash in the Los Angeles River Watershed.* [Online]. Available: http://www. waterboards.ca.gov/losangeles/board_decisions/basin_plan_amendments/technical_ documents/2001-014/01_0622_bc_Tentative%20Resolution,%20June%2022,%202001. pdf/ [July 24, 2008].

Cantin, J., J. Eyraud, and C. Fenton. 1990. Quantitative estimates of garbage generation and disposal in the U.S. maritime sectors before and after MARPOL Annex V. In *Proceedings of the Second International Conference on Marine Debris*, Shomura, R.S. and M.L. Godfrey (eds.). U.S. Department of Commerce, Washington, DC.

Carpentaria Ghost Nets Programme. 2008. *Carpentaria Ghost Nets Programme.* [Online]. Available: http://www.ghostnets.com.au/ [September 1, 2008].

Carpenter, A. and S.M. Macgill. 2001. Charging for port reception facilities in North Sea ports: Putting theory into practice. *Marine Pollution Bulletin* 42:257-266.

Carr, H.A., E.H. Amaral, A.W. Hulbert, and R. Cooper. 1985. Underwater survey of simulated lost demersal and lost commercial gill nets off New England. In *Proceedings of the Workshop on the Fate and Impact of Marine Debris*, Shomura, R.S. and H.O. Yoshida (eds.). U.S. Department of Commerce, Washington, DC.

Carretta, J.V., K.A. Forney, M.S. Lowry, J. Barlow, J. Baker, B. Hanson, and M.M. Muto. 2007. *U.S. Pacific Marine Mammal Stock Assessments: 2007.* Southwest Fisheries Science Center, National Oceanic and Atmospheric Administration, U. S. Department of Commerce, La Jolla, California.

Chiappone, M., H. Dienes, D.W. Swanson, and S.L. Miller. 2005. Impacts of lost fishing gear on coral reef sessile invertebrates in the Florida Keys National Marine Sanctuary. *Biological Conservation* 121:221-230.

City of Unalaska. 2008. *Schedule of Fees and Charges.* [Online]. Available: http:// unalaska-ak.us/vertical/Sites/%7B0227B6A7-A82F-4BFC-9D02-A4B2D3A8BC35%7D/ uploads/%7B30A5922A-CF11-4783-85B3-47D7A36E4473%7D.PDF [August 5, 2008].

Coe, J.M. 1990. A review of marine debris research, education, and mitigation in the North Pacific. In *Proceedings of the Second International Conference on Marine Debris*, Shomura, R.S. and M.L. Godfrey (eds.). U.S. Department of Commerce, Washington, DC.

Coe, J.M. and D.B. Rogers (eds.). 1997. *Marine Debris: Sources, Impacts and Solutions.* Springer, New York.

Croxall, J.P., S. Rodwell, and I.L. Boyd. 1990. Entanglement in man-made debris of Antarctic fur seals at Bird Island, South Georgia. *Marine Mammal Science* 6:221-233.

Cruise Lines International Association, Inc. 2006. *CLIA Industry Standard: Cruise Industry Waste Management Practices and Procedures.* Cruise Lines International Association, Inc., Fort Lauderdale, Florida.

Dameron, O.J., M. Parke, M.A. Albins, and R. Brainard. 2007. Marine debris accumulation in the Northwestern Hawaiian Islands: An examination of rates and processes. *Marine Pollution Bulletin* 54:423-433.

Delgado de Molina, A., J. Ariz, J.C. Santana, and S. Déniz. 2006. *Study of Alternative Models of Artificial Floating Objects for Tuna Fishery (Experimental Purse-seine Campaign in the Indian Ocean).* Indian Ocean Tuna Commission, Mahé, Seychelles.

Donohue, M.J. 2003. How multi-agency partnerships can successfully address large-scale pollution problems: A Hawaii case study. *Marine Pollution Bulletin* 46:700-702.

Donohue, M.J. 2005. Eastern Pacific Ocean source of Northwestern Hawaiian Islands marine debris supported by errant fish aggregating device. *Marine Pollution Bulletin* 50(8):886-888.

Donohue, M.J. and D.G. Foley. 2007. Remote sensing reveals links among the endangered Hawaiian monk seal, marine debris and El Niño. *Marine Mammal Science* 23(2):468-473.

Donohue, M.J., R.C. Boland, C.M. Sramek, and G.A. Antonelis. 2001. Derelict fishing gear in the Northwestern Hawaiian Islands: Diving surveys and debris removal in 1999 confirm threat to coral reef ecosystems. *Marine Pollution Bulletin* 42(12):1301-1312.

Edyvane, K.S., A. Dalgetty, P.W. Hone, J.S. Higham, and N.M. Wace. 2004. Long-term marine litter monitoring in the remote Great Australian Bight, South Australia. *Marine Pollution Bulletin* 48:1060-1075.

Endo, S., R. Takizawa, K. Okuda, H. Takada, K. Chiba, H. Kanehiro, H. Ogi, R. Yamashita, and T. Date. 2005. Concentration of polychlorinated biphenyls (PCBs) in beached resin pellets: Variability among individual particles and regional differences. *Marine Pollution Bulletin* 50:1103-1114.

Environmental Protection Agency. 1993. *Guidance Specifying Management Measures for Sources of Nonpoint Pollution of Coastal Waters*. [Online]. Available: http://www.epa.gov/nps/MMGI/ [September 9, 2008].

Environmental Protection Agency. 2002. *Assessing and Monitoring Floatable Debris*. [Online]. Available: http://www.epa.gov/owow/oceans/debris/floatingdebris/debris-final.pdf [June 13, 2008].

Environmental Protection Agency. 2003. *Shipshape Shores and Waters: A Handbook for Marina Operators and Recreational Boaters*. [Online]. Available: http://www.epa.gov/nps/marinashdbk2003.pdf [July 10, 2008].

Environmental Protection Agency. 2004. *Environmental Management Systems: Systematically Improving Your Performance*. [Online]. Available: http://www.epa.gov/ispd/shipbuilding/bizcase.pdf [July 10, 2008].

Environmental Protection Agency. 2007a. *Floatables Action Plan Assessment Report 2006*. United States Environmental Protection Agency, Region 2, Division of Environmental Science and Assessment, Washington, DC.

Environmental Protection Agency. 2007b. *Draft Cruise Ship Discharge Assessment Report*. [Online]. Available: http://www.epa.gov/owow/oceans/cruise_ships/disch_assess_draft.html [June 13, 2008].

Escardó-Boomsma, J., K. O'Hara, and C.A. Ribic. 1995. *National Marine Debris Monitoring Program, Volumes 1-2, Final Report*. Office of Water, Environmental Protection Agency, Washington, DC.

Fonteneau, A., P. Pallarés, and R. Pianet. 2000. A worldwide review of purse seine fisheries on FADs. In *Tuna Fishing and Fish Aggregating Devices*, Le Gall, J.Y., P. Cayré, and M. Taquet (eds.). Institut Français de Recherche pour l'Exploitation de la Mer (IFREMER), Plouzané, France.

Food and Agriculture Organization of the United Nations. 1995. *Code of Conduct for Responsible Fisheries*. [Online]. Available: ftp://ftp.fao.org/docrep/fao/005/v9878e/v9878e00.pdf [August 26, 2008].

Food and Agriculture Organization of the United Nations. 2008a. *Fish Capture Technology*. [Online]. Available: http://www.fao.org/fishery/topic/3384/en [July 30, 2008].

Food and Agriculture Organization of the United Nations. 2008b. *Geartype Fact Sheets: Driftnets*. [Online]. Available: http://www.fao.org/fishery/geartype/220 [July 24, 2008].

Food and Agriculture Organization of the United Nations. 2008c. *Fish Aggregating Devices*. [Online]. Available: http://www.fao.org/fishery/topic/14889/en [July 24, 2008].

Fowler, C.W. 1987. Marine debris and northern fur seals: A case study. *Marine Pollution Bulletin* 18:326-335.

Fowler, C.W. and N. Baba. 1991. *Entanglement Studies, St. Paul Island: 1990 Juvenile Male Northern Fur Seals*. Alaska Fisheries Science Center, National Marine Fisheries Service, U.S. Department of Commerce, Seattle, Washington.

Freeman, A.M., III. 1984. The quasi-option value of irreversible development. *Journal of Environmental Economics and Management* 11:292-295.

Fry, D.M., S.I. Fefer, and L. Sileo. 1987. Ingestion of plastic debris by Laysan albatrosses and wedge-tailed shearwaters in the Hawaiian Islands. *Marine Pollution Bulletin* 18:339-343.

Furness, R.W. 1985. Ingestion of plastic particles by seabirds at Gough Island, South Atlantic Ocean. *Environmental Pollution Series A: Ecological and Biological* 38:261-272.

Gates, J.E. and L.W. Gysel. 1978. Avian nest dispersion and fledging success in field–forest ecotones. *Ecology* 59:871-883.

Georgakellos, D.A. 2007. The use of the deposit-refund framework in port reception facilities charging systems. *Marine Pollution Bulletin* 54:508-520.

George, G.A. 1995. Weathering of polymers: Mechanisms of degradation and stabilization, testing strategies and modelling. *Materials Forum* 19:145-161.

Gerrodette, T., B.K. Choy, and L.M. Hiruki. 1985. An experimental study of derelict gillnet fragments in the central Pacific Ocean. In *Proceedings of the Workshop on the Fate and Impact of Marine Debris*, Shomura, R.S. and H.O. Yoshida (eds.). U.S. Department of Commerce, Washington, DC.

Gillett, R., M.A. McCoy, and D.G. Itano. 2002. *Status of the United States Western Pacific Tuna Purse Seine Fleet and Factors Affecting Its Future*. School of Ocean and Earth Science and Technology Publication 02-01, Joint Institute for Marine and Atmospheric Research Contribution 02-344.

Gochfeld, M. 1973. Effect of artifact pollution on the viability of seabird colonies on Long Island, New York. *Environmental Pollution* 4:1-6.

Gregory, M.R. 1977. Plastic pellets on New Zealand beaches. *Marine Pollution Bulletin* 8:82-84.

Gregory, M.R. 1978. Accumulation and distribution of virgin plastic granules on New Zealand beaches. *New Zealand Journal of Marine and Freshwater Research* 12:399-414.

Gregory, M.R. 1996. Plastic "scrubbers" in hand cleansers: A further (and minor) source for marine pollution identified. *Marine Pollution Bulletin* 32:867-871.

Gregory, M.R. and P.G. Ryan. 1997. Pelagic plastics and other seaborne persistent synthetic debris: A review of southern hemisphere perspectives. In *Marine Debris: Sources, Impacts and Solutions*, Coe, J.M. and D.B. Rogers (eds.). Springer, New York.

Guillory, V., A. McMillen-Jackson, L. Hartman, T.H. Perry, T. Floyd, T. Wagner, and G. Graham. 2001. *Blue Crab Derelict Traps and Trap Removal Programs*. Gulf States Marine Fisheries Commission, Ocean Springs, Mississippi.

Hall, M.A. 1994. Bycatches in purse-seine fisheries. In *By-Catches in Fisheries and Their Impact on the Ecosystem*, Pitcher, T.J. and R. Chuenpagdee (eds.). Fisheries Centre, University of British Columbia, Vancouver, Canada.

Hallier, J.P. and D. Gaertner. 2008. Drifting fish aggregation devices could act as an ecological trap for tropical tuna species. *Marine Ecology Progress Series* 353:255-264.

Hanni, K.D. and P. Pyle. 2000. Entanglement of pinnipeds in synthetic materials at Southeast Farallon Island, California, 1976-1998. *Marine Pollution Bulletin* 40:1076-1081.

Hareide, N.-R., G. Garnes, D. Rihan, M. Mulligan, P. Tyndall, M. Clark, P. Connolly, R. Misund, P. McMullen, D. Furevik, O.B. Humborstad, K. Høydal, and T. Blasdale. 2005. *A Preliminary Investigation on Shelf Edge and Deepwater Fixed Net Fisheries to the West and North of Great Britain, Ireland, around Rockall and Hatton Bank*. [Online]. Available: http://www.seafish.org/pdf.pl?file=seafish/Documents/DEEPNETfinalreport.pdf [August 29, 2008].

Helsinki Commission. 2007. *Towards a Baltic Sea with Environmentally Friendly Maritime Activities: HELCOM Overview 2007.* [Online]. Available: http://www.helcom.fi/stc/files/Krakow2007/Maritime_activities_MM2007.pdf [September 1, 2008].

Henderson, J.R. 2001. A pre- and post-MARPOL Annex V Summary of Hawaiian monk seal entanglements and marine debris accumulation in the Northwestern Hawaiian Islands, 1982-1998. *Marine Pollution Bulletin* 42(7):584-589.

Henderson, J.R. and R. Steiner. 2000. Source identification of derelict fishing gear: Issues and concerns. In *Proceedings of the International Marine Debris Conference on Derelict Fishing Gear and the Ocean Environment,* McIntosh, N., K. Simonds, M. Donohue, C. Brammer, S. Manson, and S. Carbajal (eds.). Hawaiian Islands Humpback Whale National Marine Sanctuary, U.S. Department of Commerce.

Henderson, J.R., S.L. Austin, and M.B. Pillos. 1987. *Summary of Webbing and Net Fragments Found on Northwestern Hawaiian Islands Beaches, 1982–1986.* Southwest Fisheries Science Center, National Marine Fisheries Service, La Jolla, California.

Hess, N.A., C.A. Ribic, and I. Vining. 1999. Benthic marine debris, with an emphasis on fishery-related items, surrounding Kodiak Island, Alaska, 1994–1996. *Marine Pollution Bulletin* 38:885-890.

Heyerdahl, T. 1970. Conservation around the world. *Biological Conservation* 2:221-222.

High, W.L. 1985. Some consequences of lost fishing gear. In *Proceedings of the Workshop on the Fate and Impact of Marine Debris,* Shomura, R.S. and H.O. Yoshida (eds.). U.S. Department of Commerce, Washington, DC.

Hofmeyr, G.J.G. and M.N. Bester. 2002. Entanglement of pinnipeds at Marion Island. *South African Journal of Marine Science* 24:383-386.

Hofmeyr, G.J.G., M.N. Bester, S.P. Kirkman, C. Lydersen, and K.M. Kovacs. 2006. Entanglement of Antarctic fur seals at Bouvetøya, Southern Ocean. *Marine Pollution Bulletin* 52:1077-1080.

Hollstrom, A. 1975. Plastic films on the bottom of the Skagerrak. *Nature* 255:622-623.

Indian Ocean Tuna Commission. 2007. *Report of the Ninth Session of the IOTC Working Party on Tropical Tunas.* Indian Ocean Tuna Commission, Mahé, Seychelles.

Ingraham, W.J., Jr. and C.C. Ebbesmeyer. 2001. Surface current concentration of floating marine debris in the North Pacific Ocean: Twelve-year OSCURS model experiments. In *Proceedings of the International Conference on Derelict Fishing Gear and the Ocean Environment,* McIntosh, N., K. Simonds, M. Donohue, C. Brammer, S. Manson, and S. Carbajal (eds.). Hawaiian Islands Humpback Whale National Marine Sanctuary, U.S. Department of Commerce.

Inter-American Tropical Tuna Commission. 1999. *Report of the Chair of the Working Group on Fish-Aggregating Devices.* [Online]. Available: http://www.iattc.org/PDFFiles/fish-aggregating%20devices%202nd%20meeting%20ENG.pdf [July 31, 2008].

Inter-American Tropical Tuna Commission. 2003. *Convention for the Strengthening of the Inter-American Tropical Tuna Commission Established by the 1949 Convention between the United States of America and the Republic of Costa Rica.* [Online]. Available: http://www.iattc.org/PDFFiles2/Antigua_Convention_Jun_2003.pdf [August 28, 2008].

Inter-American Tropical Tuna Commission. 2007a. *The Fishery for Tunas and Billfishes in the Eastern Pacific Ocean in 2006.* Inter-American Tropical Tuna Commission, La Jolla, California.

Inter-American Tropical Tuna Commission. 2007b. *Annual Report of the Inter-American Tropical Tuna Commission, 2005.* Inter-American Tropical Tuna Commission, La Jolla, California.

International Commission for the Conservation of Atlantic Tunas. 2001. Report of the evaluation of the effect of the time/area closure of fishing under objects of the surface fleets. *Collective Volume of Scientific Papers ICCAT* 52(2):350-414.

International Commission for the Conservation of Atlantic Tunas. 2003. *Recommendation by ICCAT Concerning the Duties of Contracting Parties and Cooperating Non-contracting Parties, Entities, or Fishing Entities in Relation to Their Vessels Fishing in the ICCAT Convention Area.* [Online]. Available: http://www.iccat.int/Documents/Recs/compendiopdf-e/2003-12-e.pdf [September 3, 2008].

International Commission for the Conservation of Atlantic Tunas. 2004. *Recommendation by ICCAT on a Multi-year Conservation and Management Program from Bigeye Tuna.* [Online]. Available http://www.iccat.int/Documents/Recs/compendiopdf-e/2004-01-e.pdf [September 11, 2008].

International Maritime Organization. 2006a. *Marine Environment Protection Committee (MEPC), 55th session: 9-13 October 2006.* [Online]. Available: http://www.imo.org/Newsroom/mainframe.asp?topic_id=109&doc_id=6219 [July 1, 2008].

International Maritime Organization. 2006b. *Guidelines for the Implementation of Annex V of MARPOL.* International Maritime Organization, London.

International Maritime Organization. 2006c. *Annex V of MARPOL 73/78: Regulations for the Prevention of Pollution by Garbage from Ships.* Consolidated Edition, International Maritime Organization, London.

International Maritime Organization. 2006d. *Guidelines on the Convention on the Prevention of Marine Pollution by Dumping of Wastes and Other Matter, 1972.* International Maritime Organization, London.

International Maritime Organization. 2007. *Interpretations and Amendments of MARPOL and Related Instruments: Report of the Correspondence Group for the Review of MARPOL Annex V.* Marine Environment Protection Committee, International Maritime Organization, London.

International Maritime Organization. 2008a. *International Maritime Organization.* [Online]. Available: http://www.imo.org/About/mainframe.asp?topic_id=3 [June 13, 2008].

International Maritime Organization. 2008b. *Status of Summary of Conventions.* [Online]. Available: http://www.imo.org/Conventions/mainframe.asp?topic_id=247 [July 29, 2008].

International Maritime Organization. 2008c. *Global Integrated Shipping Information System.* [Online]. Available: http://gisis.imo.org/Public/ [July 29, 2008].

Itano, D.G. 2007a. *An Examination of FAD-Related Gear and Fishing Strategies Useful for Data Collection and FAD-Based Management.* Western and Central Pacific Fisheries Commission, Kolonia, Pohnpei State, Federated States of Micronesia.

Itano, D.G. 2007b. *A Summary of Operational, Technical and Fishery Information on WCPO Purse Seine Fisheries Operating on Floating Objects.* Western and Central Pacific Fisheries Commission, Kolonia, Pohnpei State, Federated States of Micronesia.

Itano, D., S. Fukofuka, and D. Brogan. 2004. *The Development, Design and Recent Status of Anchored and Drifting FADs in the WCPO.* 17th Meeting of the Standing Committee on Tuna and Billfish.

Johnson, S.W. 1994. Deposition of trawl web on an Alaska beach after implementation of MARPOL Annex V legislation. *Marine Pollution Bulletin* 28:477-481.

Johnson, S.W. and J.H. Eiler. 1999. Fate of radio-tagged trawl web on an Alaskan beach. *Marine Pollution Bulletin* 38:136-141.

Joint Ocean Commission Initiative. 2007. *An Agenda for Action: Moving Regional Ocean Governance from Theory to Practice.* [Online]. Available: http://www.jointoceancommission.org/resource-center/1-Reports/2007-08-01_Agenda_for_Action_Regional_Ocean_Governance.pdf [August 27, 2008].

Kaoru, Y., V.K. Smith, and J.L. Liu. 1995. Using random utility models to estimate the recreation value of estuarine resources. *American Journal of Agricultural Economics* 77:141-151.

Kappenman, K.M. and B.L. Parker. 2007. Ghost nets in the Columbia River: Methods for locating and removing derelict gill nets in a large river and an assessment of impact to white sturgeon. *North American Journal of Fisheries Management* 27:804-809.

Keeney, R.L. and H. Raiffa. 1976. *Decisions with Multiple Objectives: Preferences and Value Tradeoffs.* Wiley, New York.

Kiessling, I. 2003. *Finding Solutions: Derelict Fishing Gear and Other Marine Debris in Northern Australia.* National Oceans Office, Hobart, Tasmania, Australia.

Kirkley, J. and K.E. McConnell. 1997. Marine debris: Benefits, costs, and choices. In *Marine Debris: Sources, Impacts and Solutions*, Coe, J.M. and D.B. Rogers (eds.). Springer, New York.

Kock, K.-H. 2001. The direct influence of fishing and fishery-related activities on non-target species in the Southern Ocean with particular emphasis on longline fishing and its impact on albatrosses and petrels—a review. *Reviews in Fish Biology and Fisheries* 11(1):31-56.

Koehler, H.R., B. Stewart, P. Carroll, and T. Rice. 2000. Legal instruments for the prevention and management of disposal and loss of fishing gear at sea. In *Proceedings of the International Marine Debris Conference on Derelict Fishing Gear and the Ocean Environment*, McIntosh, N., K. Simonds, M. Donohue, C. Brammer, S. Manson, and S. Carbajal (eds.). Hawaiian Islands Humpback Whale National Marine Sanctuary, U.S. Department of Commerce.

Kristjonsson, H. (ed.) 1959. *Modern Fishing Gear of the World.* Fishing News Limited, London.

Kubota, M. 1994. A mechanism for the accumulation of floating marine debris north of Hawaii. *Journal of Physical Oceanography* 24:1059-1064.

Kubota, M., K. Takayama, and D. Namimoto. 2005. Pleading for the use of biodegradable polymers in favor of marine environments and to avoid an asbestos-like problem for the future. *Applied Microbiology and Biotechnology* 67:469-476.

Laist, D.W. 1997. Impact of marine debris: Entanglement of marine life in marine debris including a comprehensive list of species with entanglement and ingestion records. In *Marine Debris: Sources, Impacts and Solutions*, Coe, J.M. and D.B. Rogers (eds.). Springer, New York.

Lewis, P.N., M.J. Riddle, and S.D.A. Smith. 2005. Assisted passage or passive drift: A comparison of alternative transport mechanisms for non-indigenous coastal species into the Southern Ocean. *Antarctic Science* 17:183-191.

Lovett, G.M., D.A. Burns, C.T. Driscoll, J.C. Jenkins, M.J. Mitchell, L. Rustad, J.B. Shanley, G.E. Likens, and R. Haeuber. 2007. Who needs environmental monitoring? *Frontiers in Ecology and the Environment* 5:253-260.

Lucas, Z. 1992. Monitoring persistent litter in the marine environment on Sable Island, Nova Scotia. *Marine Pollution Bulletin* 24:192-199.

Manheim, B.S., Jr. 1986. The oceans are choking on plastic debris. *New York Times,* 1 March, 26.

Marine Conservation Alliance Foundation. 2008. *Marine Debris Map.* [Online]. Available: http://www.mcafoundation.org/googlemap.html [June 23, 2008].

Mato, Y., T. Isobe, H. Takada, H. Kanehiro, C. Ohtake, and T. Kaminuma. 2001. Plastic resin pellets as a transport medium for toxic chemicals in the marine environment. *Environmental Science and Technology* 35:318-324.

Matsumura, S. and K. Nasu. 1997. Distribution of floating debris in the North Pacific Ocean: Sighting surveys 1986-1991. In *Marine Debris: Sources, Impacts and Solutions*, Coe, J.M. and D.B. Rogers (eds.). Springer, New York.

Matsuoka, T., T. Nakashima, and N. Nagasawa. 2005. A review of ghost fishing: Scientific approaches to evaluation and solutions. *Fisheries Science* 71:691-702.

McDermid, K.J. and T.L. McMullen. 2004. Quantitative analysis of small-plastic debris on beaches in the Hawaiian archipelago. *Marine Pollution Bulletin* 48:790-794.

Ménard, F., A. Fonteneau, D. Gaertner, V. Nordstrom, B. Stéquert, and E. Marchal. 2000. Exploitation of small tunas by a purse-seine fishery with fish aggregating devices and their feeding ecology in an eastern tropical Atlantic ecosystem. *ICES Journal of Marine Science* 57:525-530.

Merrell, T.R. 1980. Accumulation of plastic litter on beaches of Amchitka Island, Alaska. *Marine Environmental Research* 3:171-184.

Merrell, T.R. 1984. A decade of change in nets and plastic litter from fisheries off Alaska. *Marine Pollution Bulletin* 15:378-384.

Merrell, T.R. 1985. Fish nets and other plastic litter on Alaska beaches. In *Proceedings of the Workshop on the Fate and Impact of Marine Debris*, Shomura, R.S. and H.O. Yoshida (eds.). U.S. Department of Commerce, Washington, DC.

Miller, J.E. and E.R. Jones. 2003. *A Study of Shoreline Trash: 1989–1998 Padre Island National Seashore*. National Park Service, U.S. Department of the Interior.

Moore, C.J., S.L. Moore, M.K. Leecaster, and S.B. Weisberg. 2001a. A comparison of plastic and plankton in the North Pacific central gyre. *Marine Pollution Bulletin* 42:1297-1300.

Moore, C.J., D. Gregorio, M. Carreon, M.K. Leecaster, and S.B. Weisberg. 2001b. Composition and distribution of beach debris in Orange County, California. *Marine Pollution Bulletin* 42:241-245.

Moreno, G., L. Dagorn, G. Sancho, and D. Itano. 2007. Fish behaviour from fishers knowledge: The case study of tropical tuna around drifting fish aggregating devices (DFADs). *Canadian Journal of Fisheries and Aquatic Sciences* 64(11):1517-1528.

Morishige, C., M.J. Donahue, E. Flint, C. Swenson, and C. Woolaway. 2007. Factors affecting marine debris deposition at French Frigate Shoals, Northwestern Hawaiian Islands Marine National Monument, 1990-2006. *Marine Pollution Bulletin* 54:1162-1169.

Morón, J., J. Areso, and P. Pallarés. 2001. *Statistics and Technical Information about the Spanish Purse Seine Fleet in the Pacific*. [Online]. Available: http://www.spc.int/OceanFish/Html/SCTB/SCTB14/ftwg11.pdf [August 4, 2008].

Moser, M.L. and D.S. Lee. 1992. A fourteen-year survey of plastic ingestion by western North Atlantic seabirds. *Colonial Waterbirds* 15:83-94.

Nagelkerken, I., G.A.M.T. Wiltjer, A.O. Debrot, and L.P.J.J. Pors. 2001. Baseline study of submerged marine debris at beaches in Curaçao, West Indies. *Marine Pollution Bulletin* 42:786-789.

National Oceanic Atmospheric Administration. 2007a. *U.S. Response to Inquiry Regarding: (1) What Is Being Done to Reduce Entanglement of Large Whales in U.S. Waters, (2) What Is the Response Protocol once an Entangled Large Whale is Encountered, and (3) What Is the U.S. Policy Regarding Euthanasia of Entangled Large Whales in U.S. Waters*. [Online]. Available: http://www.iwcoffice.org/_documents/commission/IWC59docs/59-17.pdf [July 24, 2008].

National Oceanic and Atmospheric Administration. 2007b. *Fisheries of the United States, 2006*. Fisheries Statistics Division, Office of Science and Technology, National Marine Fisheries Service, Silver Spring, Maryland.

National Oceanic and Atmospheric Administration. 2008a. *2007 Report of the Secretary of Commerce to the Congress of the United States Concerning U.S. Actions Taken on Foreign Large-Scale High Seas Driftnet Fishing*. National Marine Fisheries Service, National Oceanic and Atmospheric Administration, U.S. Department of Commerce, Washington, DC.

National Oceanic and Atmospheric Administration. 2008b. *NOAA Chesapeake Bay Office: Derelict Fishing Gear Study*. [Online]. Available: http://chesapeakebay.noaa.gov/docs/DFGPfactsheet51308.pdf [September 3, 2008].

National Research Council. 1975. *Assessing Potential Ocean Pollutants*. National Academy Press, Washington, DC.

National Research Council. 1983. *Risk Assessment in the Federal Government: Managing the Process.* National Academy Press, Washington, DC.

National Research Council. 1990. *Decline of the Sea Turtles: Causes and Prevention.* National Academy Press, Washington, DC.

National Research Council. 1993. *Issues in Risk Assessment.* National Academy Press, Washington, DC.

National Research Council. 1994. *Polymer Science and Engineering: The Shifting Research Frontiers.* National Academy Press, Washington, DC.

National Research Council. 1995a. *Clean Ships, Clean Ports, Clean Oceans: Controlling Garbage and Plastic Wastes at Sea.* National Academy Press, Washington, DC.

National Research Council. 1995b. *Understanding Marine Biodiversity.* National Academy Press, Washington, DC.

National Research Council. 1996a. *Shipboard Pollution Control: U.S. Navy Compliance with MARPOL Annex V.* National Academy Press, Washington, DC.

National Research Council. 1996b. *Stemming the Tide: Controlling Introductions of Nonindigenous Species by Ships' Ballast Water.* National Academy Press, Washington, DC.

National Research Council. 1996c. *Understanding Risk: Informing Decisions in a Democratic Society.* National Academy Press, Washington, DC.

National Research Council. 1999. *Sharing the Fish: Toward a National Policy on Individual Fishing Quotas.* National Academy Press, Washington DC.

National Research Council. 2002. *Animal Biotechnology: Science-Based Concerns.* National Academy Press, Washington, DC.

National Research Council. 2003. *Decline of the Steller Sea Lion in Alaskan Waters: Untangling Food Webs and Fishing Nets.* The National Academies Press, Washington, DC.

National Research Council. 2004. *Nonnative Oysters in the Chesapeake Bay.* The National Academies Press, Washington, DC.

Natural Resources Consultants, Inc. 2007. *A Cost-Benefit Analysis of Derelict Fishing Gear Removal in Puget Sound, Washington.* Prepared for the Northwest Straits Commission, Mt. Vernon, Washington.

Natural Resources Consultants, Inc. 2008. *Rates of Marine Species Mortality Caused by Derelict Fishing Nets in Puget Sound, Washington.* Prepared for the Northwest Straits Commission, Mt. Vernon, Washington.

Northwest Straits Commission. 2008. *Derelict Fishing Gear Removal.* [Online]. Available: http://www.nwstraits.org/PageID/170/default.aspx [August 6, 2008].

Ofiara, D.D. and B. Brown. 1999. Assessment of economic losses to recreational activities from 1988 marine pollution events and assessment of economic losses from long-term contamination of fish within the New York Bight to New Jersey. *Marine Pollution Bulletin* 38(11):990-1004.

O'Hara, K.J., S. Iudicello, and R. Bierce. 1988. *A Citizens Guide to Plastics in the Ocean: More Than a Litter Problem.* Center for Environmental Education, Washington, DC.

Olson, P.H. 1994. Handling of waste in ports. *Marine Pollution Bulletin* 29:284-295.

Page, B., J. McKenzie, R. McIntosh, A. Baylis, A. Morrissey, N. Calvert, T. Haase, M. Berris, D. Dowie, P.D. Shaughnessy, and S.D. Goldsworth. 2004. Entanglement of Australian sea lions and New Zealand fur seals in lost fishing gear and other marine debris before and after government and industry attempts to reduce the problem. *Marine Pollution Bulletin* 49:33-42.

Pew Oceans Commission. 2003. *America's Living Oceans: Charting a Course for Sea Change.* Pew Oceans Commission, Arlington, Virginia.

Pichel, W.G., J.H. Churnside, T.S. Veenstra, D.G. Foley, K.S. Friedman, R.E. Brainard, J.B. Nicoll, Q. Zheng, and P. Clemente-Colón. 2007. Marine debris collects within the North Pacific Subtropical Convergence Zone. *Marine Pollution Bulletin* 54:1207-1211.

Pierce, K.E., R.J. Harris, L.S. Larned, and M.A. Pokras. 2004. Obstruction and starvation associated with plastic ingestion in a northern gannet *Morus bassnus* and a greater shearwater *Puffinus gravis*. *Marine Ornithology* 32:187-189.

Polovina, J.J., E. Howell, D.R. Kobayashi, and M.P. Seki. 2001. The transition zone chlorophyll front, a dynamic global feature defining migration and forage habitat for marine resources. *Progress in Oceanography* 29:469-483.

Pooley, S.G. 2000. Economics of lost fishing gear. In *Proceedings of the International Marine Debris Conference on Derelict Fishing Gear and the Ocean Environment*, McIntosh, N., K. Simonds, M. Donohue, C. Brammer, S. Manson, and S. Carbajal (eds.). Hawaiian Islands Humpback Whale National Marine Sanctuary, U.S. Department of Commerce.

Pratt, J.W., H. Raiffa, and R. Schlaifer. 1995. *Introduction to Statistical Decision Theory*. MIT Press, Cambridge.

Reddy, M.S., S. Basha, S. Adimurthy, and G. Ramachandraiah. 2006. Description of the small plastics fragments in marine sediments along the Alang-Sosiya ship-breaking yard, India. *Estuarine, Coastal Shelf Science* 68:656-660.

Redford, D.P., H.K. Trulli, and W.R. Trulli. 1997. Sources of plastic pellets in the aquatic environment. In *Marine Debris: Sources, Impacts and Solutions*, Coe, J.M. and D.B. Rogers (eds.). Springer, New York.

Ribic, C.A. 1990. Report of the working group on methods to assess the amount and types of marine debris. In *Proceedings of the Second International Conference on Marine Debris*, Shomura, R.S. and M.L. Godfrey (eds.). U.S. Department of Commerce, Washington, DC.

Ribic, C.A., T.R. Dixon, and I. Vining. 1992. *Marine Debris Survey Manual*. National Marine Fisheries Service, National Oceanic and Atmospheric Administration, Seattle, Washington.

Ribic, C.A., S.W. Johnson, and C.A. Cole. 1997. Distribution, type, accumulation, and source of marine debris in the United States, 1989–1993. In *Marine Debris: Sources, Impacts and Solutions*, Coe, J.M. and D.B. Rogers (eds.). Springer, New York.

Rios, L.M., C. Moore, and P.R. Jones. 2007. Persistent organic pollutants carried by synthetic polymers in the ocean environment. *Marine Pollution Bulletin* 54:1230-1237.

Robards, M.D., J.F. Piatt, and K.D. Wohl. 1995. Increasing frequency of plastic particles ingested by seabirds in the subarctic North Pacific. *Marine Pollution Bulletin* 30:151-157.

Robards, M.D., P.J. Gould, and J.F. Piatt. 1997. The highest global concentrations and increased abundance of oceanic plastic debris in the North Pacific: Evidence from seabirds. In *Marine Debris: Sources, Impacts and Solutions*, Coe, J.M. and D.B. Rogers (eds.). Springer, New York.

Robertson, B.A. and R.L. Hutto. 2006. A framework for understanding ecological traps and an evaluation of existing evidence. *Ecology* 87:1075-1085.

Ryan, P.G. 1988. Effects of ingested plastic on seabird feeding: Evidence from chickens. *Marine Pollution Bulletin* 19:125-128.

Ryan, P.G. 1990. The marine plastic debris problem off southern Africa: Types of debris, their environmental effects, and control measures. In *Proceedings of the Second International Conference on Marine Debris*, Shomura, R.S. and M.L. Godfrey (eds.). U.S. Department of Commerce, Washington, DC.

Ryan, P.G. and C.L. Moloney. 1993. Marine litter keeps increasing. *Nature* 361:23.

Ryan, P.G. and S. Jackson. 1987. The lifespan of ingested plastic particles in seabirds and their effect on digestive efficiency. *Marine Pollution Bulletin* 18:217-219.

Ryan, P.G., A.D. Connell, and B.D. Gardener. 1988. Plastic ingestion and PCBs in seabirds: Is there a relationship? *Marine Pollution Bulletin* 19:174-176.

Saaty, T.L. 1990. *Multicriteria Decision Making: The Analytic Hierarchy Process*. RWS Publications, Pittsburgh, Pennsylvania.

Sancho, G., E. Puente, A. Bilbao, E. Gomez, and L. Arregi. 2003. Catch rates of monkfish (*Lophius* spp.) by lost tangle nets in the Cantabrian Sea (northern Spain). *Fisheries Research* 64:129-139.

Santos, I.R., A.C. Friedrich, and F.P. Barretto. 2005a. Overseas garbage pollution on beaches of northeast Brazil. *Marine Pollution Bulletin* 50:783-786.

Santos, I.R., A.C. Friedrich, M. Wallner-Kersanach, and G. Fillmann. 2005b. Influence of socio-economic characteristics of beach users on litter generation. *Ocean and Coastal Management* 48:742-752.

Schlaepfer, M.A., M.C. Runge, and P.W. Sherman. 2002. Ecological and evolutionary traps. *Trends in Ecology and Evolution* 17:474-480.

Schleyer, M.H. and B.J. Tomalin. 2000. Damage on South African coral reefs and an assessment of their sustainable diving capacity using a fisheries approach. *Bulletin of Marine Science* 67:1025-1042.

Shaw, D.G. and R.H. Day. 1994. Colour- and form-dependent loss of plastic micro-debris from the North Pacific Ocean. *Marine Pollution Bulletin* 28:39-43.

Sheavly, S.B. 2007. *National Marine Debris Monitoring Program: Final Program Report, Data Analysis and Summary*. Prepared for U.S. Environmental Protection Agency by Ocean Conservancy, Washington, DC.

Shiber, J.G. 1979. Plastic pellets on the coast of Lebanon. *Marine Pollution Bulletin* 10:28-30.

Shiber, J.G. 1982. Plastic pellets on Spain's "Costa del Sol" beaches. *Marine Pollution Bulletin* 13:409-412.

Shiber, J.G. 1987. Plastic pellets and tar on Spain's Mediterranean beaches. *Marine Pollution Bulletin* 18:84-86.

Shomura, R.S. and M.L. Godfrey. (eds). 1990. *Proceedings of the Second International Conference on Marine Debris*. U.S. Department of Commerce, Washington, DC.

Sileo, L., P.R. Sievert, and M.D. Samuel. 1990. Causes of mortality of albatross chicks at Midway Atoll. *Journal of Wildlife Diseases* 26:329-338.

Smith, V.K. and R.B. Palmquist. 1994. Temporal substitution and the recreation value of coastal amenities. *Review of Economics and Statistics* 76:119-126.

Smith, V.K., X. Zhang, and R.B. Palmquist. 1997. Marine debris, beach quality, and non-market values. *Environmental and Resource Economics* 10:223-247.

Smolowitz, R.J. 1978. Trap design and ghost fishing: Discussion. *Marine Fisheries Review* 40(5-6):59-67.

Spear, L.B., D.G. Ainley, and C.A. Ribic. 1995. Incidence of plastic in seabirds from the tropical Pacific, 1984–91: Relation with distribution of species, sex, age, season, year and body weight. *Marine Environmental Research* 40:123-146.

Stevens, B.G., I. Vining, S. Byersdorfer, and W. Donaldson. 2000. Ghost fishing by Tanner crab (*Chionoecetes bairdi*) pots off Kodiak, Alaska: Pot density and catch per trap as determined from sidescan sonar and pot recovery data. *Fishery Bulletin* 98:389-399.

Swanson, R.L., T.M. Bell, J. Kahn, and J. Olha. 1991. Use impairments and ecosystem impacts of the New York Bight. *Chemistry and Ecology* 5(1):99-127.

Teuten, E.L., S.J. Rowland, T.S. Galloway, and R.C. Thompson. 2007. Potential for plastics to transport hydrophobic contaminants. *Environmental Science and Technology* 41:7759-7764.

Thompson, R.C., Y. Olsen, R.P. Mitchell, A. Davis, S.J. Rowland, A.W.G. John, D. McGonigle, and A.E. Russell. 2004. Lost at sea: Where is all the plastic? *Science* 304:838.

Timmers, M.A., C.A. Kistner, and M.J. Donohue. 2005. *Marine Debris of the Northwestern Hawaiian Islands: Ghost Net Identification*. University of Hawaii Sea Grant College Program, Honolulu, Hawaii.

Toufexis, A. 1988. The dirty seas. *TIME Magazine*, 1 August.

Uchida, R.N. 1985. The types and estimated amounts of fish net deployed in the North Pacific. In *Proceedings of the Workshop on the Fate and Impact of Marine Debris*, Shomura, R.S. and H.O. Yoshida (eds.). U.S. Department of Commerce, Washington, DC.

Uneputty, P.A. and S.M. Evans. 1997. The impact of plastic debris on the biota of tidal flats in Ambon Bay (eastern Indonesia). *Marine Environmental Research* 44:233-242.

United Nations Environment Programme. 2008. *Global Initiative.* [Online]. Available: http://www.unep.org/regionalseas/marinelitter/initiatives/unepglobal/default.asp [August 28, 2008].

Urban Harbors Institute. 2000. *America's Greenports: Environmental Management and Technology at U.S. Ports.* [Online]. Available: http://www.uhi.umb.edu/pdf_files/greenports.pdf [July 10, 2008].

U.S. Coast Guard. 2007. *Marine Debris Research, Prevention, and Reduction Act Report to Congress.* Office of Vessel Activities for the Assistant Commandant of Prevention, U.S. Coast Guard, Washington, DC.

U.S. Coast Guard. 2008. *United States Coast Guard Maritime Information Exchange: MARPOL Servicing Facilities.* [Online]. Available: http://cgmix.uscg.mil/MARPOL/Default.aspx [July 14, 2008].

U.S. Commission on Ocean Policy. 2004. *An Ocean Blueprint for the 21st Century.* Final Report. Washington, DC.

U.S. General Accounting Office. 2000. *Marine Pollution: Progress Made to Reduce Marine Pollution by Cruise Ships, but Important Issues Remain.* [Online]. Available: http://www.gao.gov/new.items/rc00048.pdf [June 12, 2008].

Van Dyke, J.M. 2008. The 1982 United Nations Convention on the Law of the Sea. In *Ocean and Coastal Law and Policy,* Baur, D.C., T. Eichenberg, and M. Sutton (eds.). ABA Publishing, Chicago.

Velander, K.A. and M. Mocogni. 1998. Maritime litter and sewage contamination at Cramond Beach Edinburgh—a comparative study. *Marine Pollution Bulletin* 36:385-389.

Vlietstra, L.S. and J.A. Parga. 2002. Long-term changes in the type, but not amount, of ingested plastic particles in short-tailed shearwaters in the southeastern Bering Sea. *Marine Pollution Bulletin* 44:945-955.

Walker, T.R., K. Reid, J.P.Y. Arnould, and J.P. Croxall. 1997. Marine debris surveys at Bird Island, South Georgia 1990–1995. *Marine Pollution Bulletin* 34:61-65.

Walsh, R.G., J.B. Loomis, and L.O. Gilliam. 1984. Valuing option, existence and bequest demands for wilderness. *Land Economics* 60:14-29.

White, D., K. Cook, and C. Hamilton. 2004. *The Net Kit: A Net Identification Guide to Northern Australia.* World Wildlife Fund, Sydney, Australia.

White House Office of the Press Secretary. 2008. *Fact Sheet: Allowing Offshore Exploration to Help Address Rising Fuel Costs.* July 14, 2008. [Online]. Available: http://www.whitehouse.gov/news/releases/2008/07/20080714-7.html [July 22, 2008].

Williams, A.T. and D.T. Tudor. 2001. Temporal trends in litter dynamics at a pebble pocket beach. *Journal of Coastal Research* 17:137-145.

Williams, P. and C. Reid. 2006. *Overview of Tuna Fisheries in the Western and Central Pacific Ocean, Including Economic Conditions—2006.* Western and Central Pacific Fisheries Commission, Kolonia, Pohnpei State, Federates States of Micronesia.

Willoughby, N.G., H. Sangkoyo, and B.O. Lakaseru. 1997. Beach litter: An increasing and changing problem for Indonesia. *Marine Pollution Bulletin* 34:469-478.

Winston, J.E., M.R. Gregory, and L.M. Stevens. 1997. Encrusters, epibionts, and other biota associated with pelagic plastics: A review of biogeographical, environmental, and conservation issues. In *Marine Debris: Sources, Impacts and Solutions,* Coe, J.M. and D.B. Rogers (eds.). Springer, New York.

Yamashita, R. and A. Tanimura. 2007. Floating plastic in the Kuroshio Current area, western North Pacific Ocean. *Marine Pollution Bulletin* 54:485-488.

Ye, S. and A.L. Andrady. 1991. Fouling of floating plastic debris under Biscayne Bay exposure conditions. *Marine Pollution Bulletin* 22:608-613.

Yoshikawa, T. and K. Asoh. 2004. Entanglement of monofilament fishing lines and coral death. *Biological Conservation* 117:557-560.

Zabin, C.J., J.T. Carlton, and L.S. Goodwin. 2004. First report of the Asian sea anemone *Diadumene lineata* from the Hawaiian Islands. *Bishop Museum Occasional Papers* 79:54-58.

Zavala-González, A. and E. Mellink. 1997. Entanglement of California sea lions, *Zalophus californianus californianus*, in fishing in the central-northern part of the Gulf of California, Mexico. *Fishery Bulletin* 95(1):180-184.

Zitko, V. and M. Hanlon. 1991. Another source of pollution by plastics: Skin cleaners with plastic scrubbers. *Marine Pollution Bulletin* 22:41-42.

Appendixes

A

Committee and Staff Biographies

COMMITTEE

Keith R. Criddle is the Ted Stevens Distinguished Professor of Marine Policy in the University of Alaska, Fairbanks Juneau Center for Fisheries and Ocean Science. Dr. Criddle earned his Ph.D. in agricultural economics from the University of California, Davis, in 1989. Dr. Criddle's research focuses on the intersection between the natural sciences and economics, especially the management of living resources. Dr. Criddle's research has explored topics ranging from the economic consequences of alternative management regimes for the governance of commercial, sport, and subsistence fisheries to the bioeconomic effects of climate change in north Pacific fisheries to the evolution of the structure of the Chilean salmon aquaculture industry in response to requirements for traceability and assurance. He has served on the North Pacific Fisheries Management Council Scientific and Statistical Committee since 1993 and as an associate editor of *Marine Resource Economics* from 1993 to 2003, and Dr. Criddle was a member of the National Research Council's (NRC's) Committee on the Introduction of Nonnative Oysters in the Chesapeake Bay, Committee to Review Individual Fishing Quotas, and Committee on the Evaluation of the Sea Grant Program Review Process. He is currently a member of the Ocean Studies Board.

Anthony F. Amos is a research fellow in the Marine Science Institute at the University of Texas. He was educated at the Glyn Technology School in Surrey, England. His oceanographic career has spanned 44 years with research expeditions to all the world's oceans and many of its seas, including 35 cruises to the Antarctic and five to the Arctic. His current research interests in Texas include studies of several aspects of nearshore and bay and estuarine processes (circulation, currents, hydrography, and tides). He has also conducted a long-term study of the barrier island beaches, including marine debris surveys on these beaches. Mr. Amos is the director of the Animal Rehabilitation Keep (ARK), which he founded in 1982. ARK rescues, rehabilitates, and releases back to the wild injured and sick sea turtles and large aquatic birds, many of which are adversely affected by marine debris and fishing gear. He is a member of the New York Academy of Science, the American Geophysical Union, the Texas Marine Mammal Stranding Network, and the Sea Turtle Stranding and Salvage Network. He is also an Honorary Lifetime Member of the Texas Marine Educators Association and has served as vice-chair of the National Science Foundation's Research Vessel Technical Enhancement Committee. He is a holder of the U.S. Antarctic Service Medal. He served on the NRC's Committee on Shipborne Waste and on various committees of the Environmental Protection Agency, the Minerals Management Service, and others regarding the marine debris problem.

Paula Carroll is a retired captain in the U.S. Coast Guard. She earned a B.S. in biology. She began her Coast Guard career in 1977. Her first assignment in the marine safety field was as Assistant Chief of the Port Operations Department at Marine Safety Office Puget Sound in Seattle. Follow-on assignments included Chief of the Waterways Management Section of the Eighth District Aids to Navigation Branch in New Orleans. In 1996, Capt. Carroll assumed command of Vessel Traffic Service Houston/Galveston for three years. From 1999 until her retirement in June 2006, she was stationed on the Fourteenth District staff in Honolulu, first as Chief of the Marine Response Branch and then ultimately as Chief of Prevention. She developed and implemented a model framework with Hawaii Sea Grant; the National Oceanic and Atmospheric Administration; The Ocean Conservancy; the U.S. Department of Defense; other federal, state, and local agencies; and nongovernmental organizations to address the Fourteenth District derelict fishing gear and marine debris impacts on coral reefs and on endangered monk seal and turtle populations. Efforts resulted in the recovery of over 180 tons of marine debris and incalculable improvement of affected marine habitat.

James M. Coe recently retired as the deputy science and research director from the Alaska Fisheries Science Center at the National Oceanic and Atmospheric Administration. Mr. Coe has a bachelors degree in zoology from the University of California, Santa Barbara; a masters degree in marine affairs from the University of Washington; and is a Ph.D. candidate in fisheries science at the University of Washington. He may be best known for his contributions to the major reduction in the incidental kill of dolphins in tuna purse seining during the 1970s; for his leadership in the movement to identify and control marine debris pollution during the mid-1980s to mid-1990s; for his role in the investigation and ultimate international ban on high seas large-scale driftnet fishing in the mid-1990s; and, finally, for his guidance and steady management of the National Oceanic and Atmospheric Administration research programs supporting marine resource management in Alaska since the late 1990s. Today, he is still considered one of the world's experts on the marine debris issue. Mr. Coe has authored more than 40 technical papers, reports, and guidelines, including a global review on marine debris. Mr. Coe retired from the National Oceanic and Atmospheric Administration in January of 2008.

Mary J. Donohue is a program specialist at the University of Hawaii Sea Grant College Program. She holds a Ph.D. and an M.A. degree from the University of California, Santa Cruz, in organismal and population biology and a B.A. degree in aquatic biology from the University of California, Santa Barbara. Dr. Donohue has been working on the issue of marine debris since 1999. Formerly with the National Oceanic and Atmospheric Administration, she administered, coordinated, and served as Chief Scientist on the first systematic expeditions to document, study, and remove marine debris from the coral reefs of the Northwestern Hawaiian Islands. Her research has been published in scientific journals, including the *Marine Pollution Bulletin*, the *Journal of Experimental Biology*, and the *Journal of Physiological and Biochemical Zoology*. She has spoken on marine debris in the United States, Japan, Canada, and the United Kingdom at international conferences, symposia, and as an invited university and public seminar speaker.

Judith Hill Harris serves as the Director of Transportation for the City of Portland, Maine. Her areas of responsibility include policy development and regulatory compliance for maritime, surface, and aviation transportation systems. Before her current position, she was the manager of fishing programs and maritime regulation for the City of Portland. She monitored not only fishing regulations but all maritime environmental issues, including ballast water, aquatic nuisance species, and air emissions. Earlier in her career, Ms. Harris worked for Saltwater Farm, a subsidiary of International Oceanographic Corporation, which was the nation's oldest and largest shipper of lobsters direct to consumers. During her tenure with Saltwater Farm, she became involved in all aspects of the lobster industry. Ms. Harris has worked as an advocate for fishermen and served on right whale ship strike and take reduction groups. She was a member of the State of Maine's Homeland Security Planning team and is the former chair of the Port of Portland's Maritime Disaster Task Force and the Commercial Fishing Vessel Safety Committee. Ms. Harris is the author or editor of a number of publications on fisheries and environmental issues. She is a current member of the Marine Board.

Kiho Kim is an associate professor and the chair of the Department of Environmental Science at American University. He received his Ph.D. in 1996 at the University of Buffalo, studying the ecology of tropical coral reefs, and did his postdoctoral work at Cornell University. His current research focuses on understanding the role of diseases in coral population ecology and the synergistic effects of environmental factors, such as nutrient pollution and ocean warming, in the decline of coral reefs. Dr. Kim has participated in working groups examining the ecology of diseases at the National Center for Ecological Analysis and Synthesis, has worked with the British Council in promoting international networking for young scientists, and is currently an advisor to the Coral Disease Working Group of the World Bank. He is a current member of the Ocean Studies Board.

Tony MacDonald is currently the director of the Urban Coast Institute at Monmouth University, West Long Branch, New Jersey. He earned a B.A. from Middlebury College and a J.D. from Fordham University. Mr. MacDonald was previously the executive director of the Coastal States Organization from 1998 to 2005. Prior to joining the Coastal States Organization, he was the special counsel and director of environmental affairs at the American Association of Port Authorities, where he represented the International Association of Ports and Harbors at the International Maritime Organization on negotiations on the London Convention. He has also practiced law with a private firm in Washington, D.C., working on environmental and legislative issues, and served as the Washington, D.C. environmental legislative representative for the Mayor of the City of New York. He specializes in environment, coastal, marine, and natural resources law and policy and federal, state, and local government affairs.

Kathy Metcalf is the Director of Maritime Affairs for the Chamber of Shipping of America, a maritime trade association which represents a significant number of U.S.-based companies that own, operate, or charter oceangoing tankers, container ships, and other merchant vessels engaged in both the domestic and international trades. She has held this position since 1997 and, in her capacity, represents maritime interests before Congress, federal and state agencies, and in international forums. This includes attending numerous sessions of the International Maritime Organization as the American shipowner representative on the U.S. delegation to the Marine Environment Protection Committee and the Maritime Safety Committee. Ms. Metcalf earned a B.S. in marine transportation and nautical sciences in 1978 from the U.S. Merchant Marine Academy and a J.D. in 1988 from the Delaware Law School. Prior to coming to the Chamber of Shipping, she served in various positions in the energy industry including deck officer aboard large oceangoing tankers, marine safety and environmental director, corporate regulatory and compliance manager, and state government affairs manager.

Alison Rieser is the Dai Ho Chun Distinguished Chair in Arts and Sciences, professor in the Department of Geography, and director of the Graduate Ocean Policy Certificate Program at the University of Hawaii at Manoa. She earned an LL.M. from Yale Law School, a J.D. cum laude from the George Washington University, and a B.S. in human ecology from Cornell University. Ms. Rieser is a specialist in marine conservation law, the role of property rights regimes in marine resource governance, and ecosystem-based approaches to fisheries management. She is currently investigating the governance structure of the Northwestern Hawaiian Islands Marine National Monument. She was a Pew Fellow in Marine Conservation from 1999 to 2002 and was professor of ocean and coastal law and Director of the Marine Law Institute at the University of Maine's School of Law from 1988 to 2006. Ms. Rieser has served on three previous NRC committees—the Committee for Review of the National Marine Fisheries Service: Use of Science and Data in Management and Litigation, the Committee to Review Individual Fishing Quotas, and the Committee on Marine Area Governance and Management.

Nina M. Young is the deputy director of external affairs at the Consortium for Oceanographic Research and Education. She is also the president of Ocean Research Conservation and Solutions Consulting. Ms. Young earned a B.A. in marine science from the Kutztown University of Pennsylvania and an M.S. in physiology (with a minor in zoology and veterinary science) from the University of Florida. In the past, she served as the director for the Marine Wildlife Conservation Program at The Ocean Conservancy. She participated in two marine debris removal cruises and led The Ocean Conservancy's effort to determine the source of debris collected from the Northwestern Hawaiian Islands.

STAFF

Susan Park is a program officer with the Ocean Studies Board. She received her Ph.D. in oceanography from the University of Delaware in

2004. Susan was a Christine Mirzayan Science and Technology Graduate Policy Fellow with the Ocean Studies Board in 2002 and joined the staff in 2006. She has worked on several reports with the National Academies, including *Nonnative Oysters in the Chesapeake Bay*, *Review of Recreational Fisheries Survey Methods*, *Dynamic Changes in Marine Ecosystems*, *A Review of the Ocean Research Priorities Plan and Implementation Strategy*, and *Genetically Engineered Organisms, Wildlife, and Habitat: A Workshop Summary*. Prior to joining the Ocean Studies Board, Susan spent time working on aquatic invasive species management with the Massachusetts Office of Coastal Zone Management and the Northeast Aquatic Nuisance Species Panel.

Jodi Bostrom is an associate program officer with the Ocean Studies Board. She earned an M.S. in environmental science from American University in 2006 and a B.S. in zoology from the University of Wisconsin-Madison in 1998. Since starting with the Ocean Studies Board in May 1999, Jodi has worked on several studies pertaining to coastal restoration, fisheries, marine mammals, nutrient overenrichment, ocean exploration, and capacity building.

B

Acronyms

APPS   Act to Prevent Pollution from Ships

CCAMLR   Commission for the Conservation of Antarctic Marine
               Living Resources
COA   Certificate of Adequacy
CWA   Clean Water Act

DFG   Derelict Fishing Gear

EEZ   Exclusive Economic Zone
EPA   Environmental Protection Agency
ESA   Endangered Species Act
ETP   Eastern Tropical Pacific

FAD   Fish Aggregating Device
FIR   Flotsam Information Record
FMC   Fishery Management Council
FMP   Fishery Management Plan

HELCOM   Helsinki Commission

IATTC   Inter-American Tropical Tuna Commission
ICCAT   International Commission for the Conservation of Atlantic
               Tunas

IMDCC	Interagency Marine Debris Coordinating Committee
IMO	International Maritime Organization
IOTC	Indian Ocean Tuna Commission
ISO	International Organization for Standardization
MARPOL	International Convention for the Prevention of Pollution from Ships, 1973, as modified by the Protocol of 1978
MDRPRA	Marine Debris Research, Prevention, and Reduction Act
MEPC	(IMO) Marine Environment Protection Committee
MERP	Marine Entanglement Research Program
MPPRCA	Marine Plastic Pollution Research and Control Act
MPRSA	Marine Protection, Research, and Sanctuaries Act
MSFCMA	Magnuson-Stevens Fishery Conservation and Management Act
NMDMP	National Marine Debris Monitoring Program
NOAA	National Oceanic and Atmospheric Administration
NPS	National Park Service
NRC	National Research Council
NWHI	Northwestern Hawaiian Islands
ODA	Ocean Dumping Act
RCRA	Resource Conservation and Recovery Act
RFMO	Regional Fisheries Management Organization
TMDL	Total Maximum Daily Load
USACE	U.S. Army Corps of Engineers
USCG	U.S. Coast Guard
WCPFC	Western and Central Pacific Fisheries Commission
WCPO	Western and Central Pacific Ocean

C

Selected Literature on Quantities and Impacts of Marine Debris

The following tables are a review of peer-reviewed published literature only. Only those studies involving numerical data (versus graphical)
were used herein; as such, this represents a selected set of data. When
replicate data were presented (e.g., by site, by time), they were summarized by the topmost treatment variable to derive means and ranges. Only
those studies reporting debris impact (Tables IV and V) with sample size
greater than 20 for a given species were included herein.

TABLE I Prevalence of Marine Debris in Coastal Environments

Location	Study Period	Number of Items (range)	
Per Unit Length of Coastline		*items • km^{-1}*	
NP, Alaska	1972–1974, 1983	390	(193–589)
WP, Indonesia	1985, 1995	6,452	(100–29,000)
NA, Gulf of Maine	1987		
MED, Israel	1988–1989	7,354	(5,834–9,176)
MED, N/E	1988–1989	32,420	(6,000–231,000)
MED, Israel	1990–1991	10,247	(5,120–27,774)
CAR, Panama	1990–1991	4	(2–2)
SP, Ducie Atoll	1991	395	
CAR, S–E	1991–1992		(1,900–11,200)
IND, Australia	1991–2000		
CAR, Curaçao	1992–1993	48,500	(8,220–88,840)
SA, Scotia Arc Isl.	1993–1997	146	(0–285)
IO, South Africa	1994–1995	37,400	(19,600–72,500)
SA, Brazil	1995	7,400	(300–60700)
SP, N Australia, GBR	1996	625	(62–1,715)
IO, N Australia	1996–1997	92	(52 – 132)
NA, Scotland	1996–1998	53	(5–235)
NA, Wales	1998–2000	2,274	(170–16,030)
SA, Brazil	2001–2004	9	(6–12)
IO, Gulf of Oman	2002	1,790	(430–6,010)
NA + SA	2002	1,080	(0–8,800)
MED, Balearic Island	2005	34,640	(30,500–39,750)
Per Unit Area of Shore		*items • km^{-2}*	
SP, Pitcairn Isl.	1991–1993	230,000	(120,000–350,000)
NP, Indonesia	1994	27,1000,000	(700,000–53,400,000)
SP, E Australia	1994	44,800	(29,400–59,100)
Red Sea, Jordan	1994–1995	4,026,100	(2,436,800–6,171,000)
SP, E Australia	2000	133,200	(34,000–298,000)
NP, Sea of Japan	2000		(207,000–3,410,000)
SA, Argentina	2000	62,462	(9,462–150,900)
NP, Mexico	2000	1,525,000	(1,238,800–1,829,700)
SA, Brazil	2001	137,580	(33,700–233,300)
NA, Scotland	2001–2003	989,189	(160,000–3,060,000)

NOTE: Mean values, when replicate sampling is carried out, are given with range in parentheses. Studies were separated according to sampling methods: per unit shoreline (km^{-1}) or unit area (km^{-2}). Region Legend: NA = North Atlantic; SA = South Atlantic, NP = North Pacific; WP = Western Pacific; SP = South Pacific; IO = Indian; CAR = Caribbean; MED = Mediterranean.

Weight (range)		Source
_g • km⁻¹_		
219,000	(122,000–255,000)	Merrell, 1984
		Willoughby et al., 1997
31,200	(5,500–68,200)	Podolsky, 1989
		Golik and Gertner, 1992
871,000	(51,000–3,137,000)	Gabrielides et al., 1991
		Bowman et al., 1998
70	(20–140)	Garrity and Levings, 1993
		Benton, 1991
	(8,200–154,000)	Corbin and Singh, 1993
8,000	(1,870–15,000)	Edyvane et al., 2004
3,832,000	(1,874,000–5,790,000)	Debrot et al., 1999
		Convey et al., 2002
101,000	(42,800–164,000)	Madzena and Lasiak, 1997
240,000	(4,000–1,199,000)	Wetzel et al., 2004
		Haynes, 1997
		Whiting, 1998
		Velander and Mocogni, 1999
		Williams et al., 2003
		Santos et al., 2005
27,000	(7,470–75,400)	Claereboudt, 2004
		Barnes and Milner, 2005
31,400	(25,300–41,000)	Martinez-Ribes et al., 2007
_g • km⁻²_		
		Benton, 1991
		Uneputty and Evans, 1997
		Frost and Cullen, 1997
		Abu-Hilal and Al-Najjar, 2004
		Cunningham and Wilson, 2003
	(13,440,000–21,440,000)	Kusui and Noda, 2003
		Acha et al., 2003
		Silva-Iñiguez and Fischer, 2003
		Oigman-Pszczol and Creed, 2007
		Storrier et al., 2007

TABLE II Prevalence of Marine Debris in Pelagic Environments

Location	Study Period	Sampling Technique
NP, Sargasso	1971	tow, 330 μm
NP	1972	tow, 150 μm
NA	1972	tow, 947 μm
NA, E USA	1973–1975	tow, 947 μm
NP, Alaska	1974–1975	tow, 363 μm
SA, S Africa	1977–1978	tow, 900 μm
SA, Cape Basin	1979	tow, 320 μm
NP	1985	tow, 330 μm
NA, Canada	1990	tow, 308 μm
NP, Subtropical	1999	tow, 333 μm
NP, Japan	2000–2001	tow, 330 μm
NP, Central	1972	visual, ship
MED, SW Malta	1979	visual (>15 cm), ship
NP	1985	visual (>2 cm), ship
MED, Eastern	1986	visual (>50 cm), ship
NA, Gulf of Mexico	1988–1989	visual, aerial
NA, Canada	1990	visual, ship
NA, Gulf of Mexico	1992–1994	visual, aerial
NP, Japan	2000	visual (>5 cm), ship
MED, Liguria	1997–2000	visual
SP, Chile	2002	visual
NP, SCZ	2005	visual, aerial

NOTE: Mean values, when replicate sampling was carried out, are given with range in parentheses. Studies were separated according to sampling methods: visual surveys or net sampling. Region Legend: NA = North Atlantic; SA = South Atlantic, NP = North Pacific; SP = South Pacific; MED = Mediterranean.

Number of Items (range) (km^{-2})		Weight (range) (g • km^{-2})		Source
3,537	(47–12,080)	287	(0.61–770)	Carpenter and Smith, 1972
	(0–34,000)	300	(0–3,500)	Wong et al., 1974
2,842	(61–5,466)	70	(10–78)	Colton et al., 1974
44	(40–80)			van Dolah et al., 1980
111				Shaw, 1977
3,639	(0–445,860)	42	(0–10,920)	Ryan, 1988
1,874	(0–3,600)			Morris, 1980a
33,183	(80–96,100)	420	(3–1,210)	Day and Shaw, 1987
23,468	(0–108,800)	3.5	(0–23)	Dufault and Whitehead, 1994
332,556		5,114		Moore et al., 2001
174,355	(0–3,520,000)	3,600	(0–153,000)	Yamashita and Tanimura, 2007
4.2				Venrick et al., 1973
2,000				Morris, 1980b
1.0			(0.2–1.8)	Day and Shaw, 1987
		12,000		McCoy, 1988
31,500	(7,700–77,200)			Lecke-Mitchell and Mullin, 1992
25	(0–112)			Dufault and Whitehead, 1994
1.2			(0.6–2.4)	Lecke-Mitchell and Mullin, 1992
0.4			(0.1–0.7)	Shiomoto and Kameda, 2005
8.8			(3.4–14.2)	Aliani et al., 2003
11	(0–54)			Thiel et al., 2003
209	(93–291)			Pichel et al., 2007

TABLE III Prevalence of Marine Debris in Benthic Environments

Location	Study Period	Sampling Technique	Depth (m)	Number of Items (km^{-2})	(range)
NA, Bay of Biscay	1992–1993	t/55 mm	0–100	204	(26–494)
NP, MED	1992–1998	t/10, 20 mm		373	(72–1,935)
MED, Eastern	1993	t/10 mm		2,330	(0–8,504)
MED, NW	1994	t/10–100 mm	<500	1,935	(35–33,237)
NP, Indonesia	1994–1995	seine		320,000	(50,000–690,000)
MED, FRA	1994–1995	t/10 mm	100–1,630		(23–7,700)
NP, NCL	1994–1995		0–10	363,428	(90,000–660,000)
NP, USA–AK	1994–1996	t/37 mm		38	(34–37)
NP, KOR	1996, 2005	t/60–65 mm		58	(0–256)
MED, GRC	1997–1998	t/150 mm	40–360	165	(89–240)
MED, Eastern	2000–2003	t/150 mm	15–320	180	(72–437)
NP, USA–NWHI	1999	snorkel	<10	31	(3–62)
NP, USA–NWHI	2000–2001	snorkel	<10	67	(16–165)
NA, USA–FL	2001	scuba	1–7	11,825	(9,844–28,750)
SA, BRA	2001	snorkel	subtidal	29,000	
MED, GRC	2003	scuba	0–25	14,900	(0–251,300)
NA, USA	2004–2005	scuba	16–20	3,975	(400–9,700)

NOTE: Sampling methods include trawl (t) of various mesh sizes and visual surveys using snorkeler or scuba divers. Materials Legend: f = fishing gear; m = metal; n = nets; p = plastics; r = rubber. Region Legend: NA = North Atlantic; SA = South Atlantic, NP = North Pacific; MED = Mediterranean.

Weight (range) (g • km^{-2})		Debris Types	Source
		p	Galgani et al., 1995a
		p	Galgani et al., 2000
			Galil et al., 1995
		p	Galgani et al., 1995b
			Uneputty and Evans, 1997
0.5		p	Galgani et al., 1996
133,285	(0–739,000)	m, p	Nagelkerken et al., 2001
		m, p	Hess et al., 1999
		f	Lee et al., 2006
		m, p	Stefatos et al., 1999
18,000	(5,000–47,000)	f, m, p	Koutsodendris et al., 2008
		n	Donohue et al., 2001
		n	Boland and Donohue, 2003
		f	Chiappone et al., 2004
		p	Oigman-Pszczol and Creed, 2007
		m, p, r	Katsanevakis and Katsarou, 2004
		f	Bauer et al., 2008

TABLE IV Marine Debris Ingestion

Common Name	Species	Region
Green Turtle	*Chelonia mydas*	NA, USA
Green Turtle	*Chelonia mydas*	NA, Gulf of Mexico
Green Turtle	*Chelonia mydas*	SA, S Brazil
Hawksbill Turtle	*Eretmochelys imbricata*	CAR, Costa Rica
Leatherback Turtle	*Dermochelys coriacea*	NA, Peru
Loggerhead Turtle	*Carreta caretta*	NA, Gulf of Mexico
Loggerhead Turtle	*Carreta caretta*	MED, Central
Loggerhead Turtle	*Carreta carreta*	NA, USA–Florida
Loggerhead Turtle	*Carreta carreta*	NA, NE Spain
Grey-Headed Albatross	*Diomedea chrysostoma*	SA, Southern Ocean
Wandering Albatross	*Diomedea exulans*	SA, Southern Ocean
Laysan Albatross	*Diomedea immutabilis*	NP, INWR
Laysan Albatross	*Diomedea immutabilis*	NP, Midway
Sooty Albatross	*Phoebetria fusca*	SA, Southern Ocean
Yellow-Nosed Albatross	*Thalassarche chlororhynchos*	SA, Southern Ocean
Giant Petrel	*Macronectes giganteus*	SA, Argentina
Giant Petrel	*Macronectes giganteus*	SA, Southern Ocean
Northern Giant Petrel	*Macronectes halli*	SA, Southern Ocean
Northern Fulmar	*Fulmarus glacialis*	ARC, Canada
Cape Petrel	*Daption capense*	SP, Antarctic
Northern Fulmar	*Fulmarus glacialis*	NA, E USA
Northern Fulmar	*Fulmarus glacialis*	NA, North Sea
Southern Fulmar	*Fulmarus glacialoides*	SP, Antarctic
Snow Petrel	*Pagodroma nivea*	SA, Southern Ocean
Snow Petrel	*Pagodroma nivea*	SP, Antarctic
Antarctic Petrel	*Thalassoica antartica*	SA, Southern Ocean
Blue Petrel	*Halobaena caerulea*	SA, Southern Ocean
Thin-Billed Prion	*Pachyptila belcheri*	SA, Southern Ocean
Antarctic Prion	*Pachyptila desolata*	SA, Southern Ocean
Salvin's Prion	*Pachyptila salvini*	SA, Southern Ocean
Broad-Billed Prion	*Pachyptila vittata*	SA, Gough Island
Broad-Billed Prion	*Pachyptila vittata*	SA, Southern Ocean
Bulwer's Petrel	*Bulweria bulwerii*	Tropical Pacific
White-Chinned Petrel	*Procellaria aequinoctialis*	SA, Southern Ocean
Cory's Shearwater	*Calonectris diomedea*	NA, E USA
Great Shearwater	*Puffinus gravis*	NA, E USA
Great Shearwater	*Puffinus gravis*	SA, Southern Ocean

Study Period	Sampling (N)	Prevalence (%)	Debris Type	Source
1988	dead (43)	56	m, r, u	Bjorndal et al., 1994
1995–1999	capture (142)	7	b, u	Seminoff et al., 2002
1997–1998	dead (38)	61	b, c, r	Bugoni et al., 2001
1970–1972	dead (29)	14		Carr and Stancyk, 1975
1980	dead (140)	13	u	Fritt, 1982
1986–1988	dead (82)	51	b	Plotkin et al., 1993
1986	bycatch (99)	8	u	Gramentz, 1988
1997	capture (241)	15	u	Witherington, 2002
NA	bycatch (54)	80	f, n, u	Tomás et al., 2002
1975–1985	capture + dead (170)	0.6	r, t	Ryan, 1987
1975–1985	capture + dead (156)	5	r, t	Ryan, 1987
1966	dead (100)	76	b, p	Kenyon and Kridler, 1969
1982–1983	capture + dead (50)	90	p	Fry et al., 1987
1979–1985	capture + dead (73)	1	r, t	Ryan, 1987
1979–1985	capture + dead (87)	2	r, t	Ryan, 1987
2002	capture (73)	73	p, r	Copello and Quintana, 2003
1979–1985	capture + dead (123)	7	r, t	Ryan, 1987
1979–1985	capture + dead (42)	7	r, t	Ryan, 1987
2002	bycatch (42)	36	u	Mallory et al., 2006
1984–1987	capture (30)	33		van Franeker, 1985a
1975–1989	capture (44)	86		Moser and Lee, 1992
1982–1984	dead (65)	92		van Franeker, 1985b
1984–1987	capture (27)	7		van Franeker and Bell, 1985
1979–1985	capture + dead (22)	5	r, t	Ryan, 1987
1984–1987	capture (27)	4		van Franeker and Bell, 1985
1975–1985	capture + dead (30)	7	r, t	Ryan, 1987
1975–1985	capture + dead (74)	92	p, t	Ryan, 1987
1979–1985	capture + dead (32)	69	p, t	Ryan, 1987
1979–1985	capture + dead (88)	59	p, t	Ryan, 1987
1979–1985	capture + dead (31)	52	p, t	Ryan, 1987
1983	capture + dead (31)	39	p	Furness, 1985a
1979–1985	capture + dead (137)	30	p, r, t	Ryan, 1987
1984–1991	capture (39)	0	p	Spear et al., 1995
1979–1985	capture + dead (201)	57	p, r, t	Ryan, 1987
1975–1989	capture (147)	25		Moser and Lee, 1992
1975–1989	capture (55)	64		Moser and Lee, 1992
1979–1985	capture + dead (50)	90	p, t	Ryan, 1987

continued

TABLE IV Continued

Common Name	Species	Region
Sooty Shearwater	*Puffinus griseus*	SA, Southern Ocean
Sooty Shearwater	*Puffinus griseus*	Tropical Pacific
Sooty Shearwater	*Puffinus griseus*	NP, US and Canada
Wedge-Tailed Shearwater	*Puffinus pacificus*	NP, Midway
Wedge-Tailed Shearwater	*Puffinus pacificus*	Tropical Pacific
Short-Tailed Shearwater	*Puffinus tenuirostris*	NP, Bering Sea
Audubon Shearwater	*Puffinus therminieri*	NA, E USA
Kerguelen Petrel	*Pterodroma brevirostris*	SA, Southern Ocean
Cook's Petrel	*Pterodroma cookii*	SP, New Zealand
Juan Fernandez Petrel	*Pterodroma externa*	Tropical Pacific
Black-Capped Petrel	*Pterodroma hasitata*	NA, E USA
Herald Petrel	*Pterodroma heraldica*	SP, Pitcairn Island
White-Winged Petrel	*Pterodroma leucoptera*	Tropical Pacific
Stejneger's Petrel	*Pterodroma longisrostris*	Tropical Pacific
Kermadec Petrel	*Pterodroma neglecta*	SP, Pitcairn Island
Black-Winged Petrel	*Pterodroma nigripennis*	Tropical Pacific
Tahiti Petrel	*Pterodroma rostrata*	Tropical Pacific
Murphy's Petrel	*Pterodroma ultima*	SP, Pitcairn Island
Storm Petrel	*Hydrobates pelagicus*	NA, Scotland
Fork-Tailed Storm Petrel	*Oceanodroma furcata*	NP, Alaska
Leach's Storm Petrel	*Oceanodroma leucoroha*	Tropical Pacific
Leach's Storm Petrel	*Oceanodroma leucoroha*	NP, Alaska
Wilson's Storm Petrel	*Oceanodroma oceanicus*	NA, E USA
Wedge-Rumped Storm Petrel	*Oceanodroma tethys*	Tropical Pacific
White-Faced Storm Petrel	*Pelagodroma marina*	SA, Southern Ocean
Common Diving Petrel	*Pelecanoides urinatrix*	SA, Southern Ocean
Crested Auklet	*Aethia crustatella*	NP, Alaska
Parakeet Auklet	*Aethia psittacula*	NP, Alaska
Whiskered Auklet	*Aethia pygmaea*	NP, Alaska
Cassin's Auklet	*Ptychoramphus aleuticus*	NP, Alaska
Marbled Murrelet	*Brachyramphus marmoratus*	NP, Alaska
Ancient Murrelet	*Synthliboramphus antiquus*	NP, Alaska
Pigeon Guillemot	*Cepphus columba*	NP, Alaska
Common Guillemot	*Uria aalge*	NP, Alaska
Thick-Billed Guillemot	*Uria lomvia*	NP, Alaska

Study Period	Sampling (N)	Prevalence (%)	Debris Type	Source
1979–1985	capture + dead (63)	51	p, r, t	Ryan, 1987
1984–1991	capture (36)	75	p, t	Spear et al., 1995
1988–1990	by–catch (20)	75		Blight and Burger, 1997
1982–1983	capture (20)	60	p	Fry et al., 1987
1984–1991	capture (85)	20	p, t	Spear et al., 1995
1997–2001	capture (330)	84	p, t	Vlietstra and Parga, 2002
1975–1989	capture (119)	5		Moser and Lee, 1992
1979–1985	capture + dead (63)	24	p, r, t	Ryan, 1987
1972–1977	capture + dead (55)	38	t	Imber, 1996
1984–1991	capture (183)	0.6	t	Spear et al., 1995
1975–1989	capture (57)	2		Moser and Lee, 1992
1991	capture + dead (29)	0		Imber et al., 1995
1984–1991	capture (110)	12	p	Spear et al., 1995
1984–1991	capture (46)	74	p, t	Spear et al., 1995
1991	capture + dead (27)	26		Imber et al., 1995
1984–1991	capture (66)	5	p	Spear et al., 1995
1984–1991	capture (121)	0.8	p	Spear et al., 1995
1991	capture + dead (37)	43		Imber et al., 1995
1983	capture (21)	0		Furness, 1985b
1988–1990	capture (21	86	p, t	Robards et al., 1995
1984–1991	capture (354)	20	p, t	Spear et al., 1995
1988–1990	capture (64)	48	p, t	Robards et al., 1995
1975–1989	capture (133)	38		Moser and Lee, 1992
1984–1991	capture (296)	0.3	p	Spear et al., 1995
1979–1985	capture + dead (24)	88	p, t	Ryan, 1987
1979–1985	capture + dead (53)	2	r, t	Ryan, 1987
1988–1990	capture (40)	3		Robards et al., 1995
1988–1990	capture (208)	94	p, t	Robards et al., 1995
1988–1990	capture (22)	0		Robards et al., 1995
1988–1990	capture (35)	11		Robards et al., 1995
1988–1990	capture (96)	0		Robards et al., 1995
1988–1990	capture (68)	0		Robards et al., 1995
1988–1990	capture (43)	2.3		Robards et al., 1995
1988–1990	capture (134)	0.8		Robards et al., 1995
1988–1990	capture (92)	0		Robards et al., 1995

continued

TABLE IV Continued

Common Name	Species	Region
Tufted Puffin	*Fratercula cirrhata*	NP, Alaska
Horned Puffin	*Fratercula corniculata*	NP, Alaska
Red Phalarope	*Phalaropus fulicaria*	NA, E USA
Red-Necked Phalarope	*Phalaropus lobatus*	NA, E USA
Kelp Gull	*Larus dominicanus*	SA, Southern Ocean
Glaucous-Winged Gull	*Larus glaucescens*	NP, Alaska
Bonaparte's Gull	*Larus philadelphia*	NA, E USA
Bridled Tern	*Sterna anaethetus*	NA, E USA
Sooty Tern	*Sterna fuscata*	Tropical Pacific
Arctic Tern	*Sterna hirundo*	SA, Southern Ocean
Black-Legged Kittiwake	*Rissa tridactyla*	NA, E USA
Black-Legged Kittiwake	*Rissa tridactyla*	NP, Alaska
Subantarctic Skua	*Catharacta antarctica*	SA, Southern Ocean
Pomerine Jaeger	*Stercorarius pomarinus*	NA, E. USA
Cape Cormorant	*Phalacrocorax capensis*	SA, Southern Ocean
Crowned Cormorant	*Phalacrocorax coronatus*	SA, Southern Ocean
Bank Cormorant	*Phalacrocorax neglectus*	SA, Southern Ocean
King Penguin	*Aptenodytes patagonicus*	SA, Southern Ocean
Rockhopper Penguin	*Eudyptes chrysocome*	SA, Southern Ocean
Macaroni Penguin	*Eudyptes chrysolophus*	SA, Southern Ocean
Gentoo Penguin	*Pygoscelis papua*	SA, Southern Ocean
Jackass Penguin	*Spheniscus demersus*	SA, Southern Ocean
Humboldt Penguin	*Spheniscus humboldti*	SA, Southern Ocean
Magellanic Penguin	*Spheniscus magellanicus*	SA, Southern Ocean

NOTE: Studies in which samples size was greater than 20 per species examined. Materials Legend: b = plastic bags; c = cloth; f = foamed plastic; m = metal; n = fishing net; p = packing plastic straps; r = rope/string/filament; t = plastic pellets; u = user plastics. Region Legend: NA = North Atlantic; SA = South Atlantic, NP = North Pacific; SP = South Pacific; CAR = Caribbean; ARC = Arctic.

Study Period	Sampling (N)	Prevalence (%)	Debris Type	Source
1988–1990	capture (489)	25	p, t	Robards et al., 1995
1988–1990	capture (120)	37	p, t	Robards et al., 1995
1975–1989	capture (55)	69		Moser and Lee, 1992
1975–1989	capture (36)	19		Moser and Lee, 1992
1979–1985	capture + dead (52)	13	p, r, t	Ryan, 1987
1988–1990	capture (21)	0		Robards et al., 1995
1975–1989	capture (32)	19		Moser and Lee, 1992
1975–1989	capture (67)	2		Moser and Lee, 1992
1984–1991	capture (64)	2	t	Spear et al., 1995
1979–1985	capture + dead (21)	0		Ryan, 1987
1975–1989	capture (41)	10		Moser and Lee, 1992
1988–1990	capture (256)	8	p, t	Robards et al., 1995
1979–1985	capture + dead (494)	23	p, r, t	Ryan, 1987
1975–1989	capture (40)	5		Moser and Lee, 1992
1979–1985	capture + dead (239)	0		Ryan, 1987
1979–1985	capture + dead (24)	0		Ryan, 1987
1979–1985	capture + dead (167)	0.6		Ryan, 1987
1979–1985	capture + dead (150)	0		Ryan, 1987
1979–1985	capture + dead (177)	1	p, r, t	Ryan, 1987
1979–1985	capture + dead (46)	0		Ryan, 1987
1979–1985	capture + dead (214)	0		Ryan, 1987
1979–1985	capture + dead (210)	0		Ryan, 1987
1979–1985	capture + dead (30)	0		Ryan, 1987
1979–1985	capture + dead (35)	0		Ryan, 1987

TABLE V Entanglement Rates among Marine Mammals

Common Name	Species Name	Region	Study Period
Antarctic fur seals	*Arctocephalus gazella*	SA, Bouetoya	1996–2002
Antarctic fur seals	*Arctocephalus gazella*	SA, Bird Island	1988–1989
Antarctic fur seals	*Arctocephalus gazella*	SI, Marion Island	1991–1996
Australian fur seal	*Arctocephalus pusillus*	SP, Tasmania	1989–1993
Cape fur seals	*Arctocephalus pusillus*	SA, SW Africa	1977–1979
New Zealand fur seals	*Arctocephalus forsteri*	SP, New Zealand	1995–2005
New Zealand fur seals	*Arctocephalus forsteri*	SP, New Zealand	1989–2002
Harbor seals	*Phoca vitulina*	NP, USA	1984–1986
Hawaiian monk seals	*Monachus schauinlandi*	NP, NWHI	1982–1998
Monk seals	*Monachus schauinlandi*	NP, subtropical	1998–2004
Elephant seals	*Mirounga leonina*	SA, Argentina	1995–2005
Northern elephant seals	*Mirounga angustirostris*	NP, USA	1984–1986
Australian sea lions	*Neophoca cinerea*	SP, Australia	1988–2002
California sea lions	*Zalophus californianus*	NP, Mexico	1991–1995
California sea lions	*Zalophus californianus*	NP, Mexico	1992
California sea lions	*Zalophus californianus*	NP, USA	1984–1986

NOTE: Entanglement rate given as percentage of population affected. Temporal Trends: (+) increasing; (–) decreasing; (0) no change; (+/–) increase followed by decrease. Materials Legend: n = nets; p = packing plastic straps; r = rope/string/monofilament. Region Legend: SA = South Atlantic; NP = North Pacific; SP = South Pacific; SI = South Indian.

Entanglement (percentage of population)	Debris Type	Temporal Trend	Source
0.02–0.06	n, p, r	–	Hofmeyr et al., 2006
0.4–1.0	n, p, r		Croxall et al., 1990
0.01–0.15	n, p, r	+	Hofmeyr and Bester, 2002
1.3–1.9	n, p, r		Pemberton et al., 1992
0.11–0.66	n, p, r	0	Shaughnessy, 1980
0.6–2.84	n, p, r	+/–	Boren et al., 2006
0.4–0.9	n, p, r	+	Page et al., 2004
0–0.11	p	+	Stewart and Yochem, 1987
0.7	n, p, r	+	Henderson, 2001
0.4–0.78		+	Donohue and Foley, 2007
0.001	r		Campagna et al., 2007
0.17–0.2	n, p, r	–	Stewart and Yochem, 1987
0.2–1.3	n, p, r	+	Page et al., 2004
0.21–0.59	n, r	+	Zavala-González and Mellink, 1997
3.9–7.9	n, p, r		Harcourt et al., 1994
0.08–0.14	n, p, r	+	Stewart and Yochem, 1987

TABLE VI Ghost Fishing Rates: Gill Nets

Region	Gear Type	Depth (m)	Catch Decay Rate (d^{-1})	Gear Half-Life (d)	Gear Life[a] (d)	Source
NA, SW Wales	gillnet	12–14	-0.0190	36	158	Kaiser et al., 1996
NA, SW Wales	trammel net	12–14	-0.0207	33	145	Kaiser et al., 1996
NA, Portugal	gillnet	15–18	-0.0542	13	55	Erzini et al., 1997
NA, Portugal	trammelnet	15–18	-0.0204	34	147	Erzini et al., 1997
NA, UK	bottom gillnets	20–25	-0.0069	101	434	Revill and Dunlin, 2003
Baltic Sea	gillnet	32–40	-0.0198	35	151	Tschernij and Larsson, 2003
NA, Portugal	gillnet	65–78	-0.0402	17	75	Santos et al., 2003
MED, Turkey	gillnet: multifilament	9–14	-0.0224	31	134	Ayaz et al., 2006
MED, Turkey	gillnet: monofilament	9–14	-0.0127	55	236	Ayaz et al., 2006
NP, Japan	gillnet: artificial reef	13	-0.0104	67	288	Akiyama et al., 2007
NP, Japan	gillnet: sandy bed	13	-0.0216	32	139	Akiyama et al., 2007

NOTE: Rates of ghost fishing by experimentally deployed entanglement nets. Region Legend: NA = North Atlantic; NP = North Pacific; MED = Mediterranean.

[a]Gear half-life and life were calculated as when the catch reached 50 percent and 5 percent of peak fishing capacity, respectively.

TABLE VII Ghost Fishing Rates: Pots

Target Species	Location	Mortality (%)	Source
Dungeness Crab	NP, Canada	53	Breen, 1987
Lobster	NP, NWHI	20	Parrish and Kazama, 1992
Blue Crabs	Gulf of Mexico, USA	55	Guillory, 1998
Brown Crab	NA, Wales	100	Bullimore et al., 2001
Lobster	NA, Wales	100	Bullimore et al., 2001
Snow Crab	Gulf of St. Lawrence	95	Hébert et al., 2001
King Crab	NA, Norway	7	Godoy et al., 2003

NOTE: Rates of ghost fishing by experimentally deployed entanglement nets. Mortality rates indicate percent mortality among those caught in pots. Region Legend: NA = North Atlantic; NP = North Pacific; NWHI = Northwestern Hawaiian Islands.

REFERENCES

Abu-Hilal, A.H. and T. Al-Najjar. 2004. Litter pollution on the Jordanian shores of the Gulf of Aqaba (Red Sea). *Marine Environmental Research* 58:39-63.

Acha, E.M., H.W. Mianzan, O. Iribarne, D.A. Gagliardini, C. Lasta, and P. Daleo. 2003. The role of the Rio de la Plata bottom salinity front in accumulating debris. *Marine Pollution Bulletin* 46:197-202.

Akiyama, S., E. Saito, and T. Watanabe. 2007. Relationship between soak time and number of enmeshed animals in experimentally lost gill nets. *Fisheries Science* 73:881-888.

Aliani, S., A. Griffa, and A. Molcard. 2003. Floating debris in the Ligurian Sea, north-western Mediterranean. *Marine Pollution Bulletin* 46(9):1142-1149.

Ayaz, A., D. Acarli, U. Altinagac, U. Ozekinci, A. Kara, and O. Ozen. 2006. Ghost fishing by monofilament and multifilament gillnets in Izmir Bay, Turkey. *Fisheries Research* 79:267-271.

Barnes, D.K.A. and P. Milner. 2005. Drifting plastic and its consequences for sessile organism dispersal in the Atlantic Ocean. *Marine Biology* 146:815-825.

Bauer, L.J., M.S. Kendall, and C.F.G. Jeffrey. 2008. Incidence of marine debris and its relationships with benthic features in Gray's Reef National Marine Sanctuary, Southeast USA. *Marine Pollution Bulletin* 56(3):402-413.

Benton, T. 1991. Oceans of garbage. *Nature* 352:113-113.

Bjorndal, K.A., A.B. Bolten, and C.J. Lagueux. 1994. Ingestion of marine debris by juvenile sea turtles in coastal Florida habitats. *Marine Pollution Bulletin* 28:154-158.

Blight, L.K. and A.E. Burger. 1997. Occurrence of plastic particles in seabirds from the eastern North Pacific. *Marine Pollution Bulletin* 34:323-325.

Boland, R.C. and M.J. Donohue. 2003. Marine debris accumulation in the nearshore marine habitat of the endangered Hawaiian monk seal, *Monachus schauinslandi* 1999–2001. *Marine Pollution Bulletin* 46:1385-1394.

Boren, L.J., M. Morrissey, C.G. Muller, and N.J. Gemmell. 2006. Entanglement of New Zealand fur seals in man-made debris at Kaikoura, New Zealand. *Marine Pollution Bulletin* 52:442-446.

Bowman, D., N. Manor-Samsonov, and A. Golik. 1998. Dynamics of litter pollution on Israeli Mediterranean beaches: A budgetary, litter flux approach. *Journal of Coastal Research* 14:418-432.

Breen, P.A. 1987. Mortality of Dungeness crabs caused by lost traps in the Fraser River estuary, British Columbia. *North American Journal of Fisheries Management* 7:429-435.

Bugoni, L., L. Krause, and M.V. Petry. 2001. Marine debris and human impacts on sea turtles in southern Brazil. *Marine Pollution Bulletin* 42:1330-1334.

Bullimore, B.A., P.B. Newman, M.J. Kaiser, S.E. Gilbert, and K.M. Lock. 2001. A study of catches in a fleet of "ghost-fishing" pots. *Fishery Bulletin* 99:247-253.

Campagna, C., V. Falabella, and M. Lewis. 2007. Entanglement of southern elephant seals in squid fishing gear. *Marine Mammal Science* 23:414-418.

Carpenter, E.J. and K.L. Smith. 1972. Plastics on the Sargasso Sea surface. *Science* 175:1240-1241.

Carr, A. and S. Stancyk. 1975. Observations on the ecology and survival outlook of the hawksbill turtle. *Biological Conservation* 8:161-172.

Chiappone, M., D.W. Swanson, S.L. Miller, and H. Dienes. 2004. Spatial distribution of lost fishing gear on fished and protected offshore reefs in the Florida Keys National Marine Sanctuary. *Caribbean Journal of Science* 40:312-326.

Claereboudt, M.R. 2004. Shore litter along sandy beaches of the Gulf of Oman. *Marine Pollution Bulletin* 49:770-777.

Colton, J.B., F.D. Knapp, and B.R. Burns. 1974. Plastic particles in surface waters of the northwestern Atlantic. *Science* 185:491-497.

Convey, P., D.K.A. Barnes, and A. Morton. 2002. Debris accumulation on oceanic island shores of the Scotia Arc, Antarctica. *Polar Biology* 25:612-617.

Copello, S. and F. Quintana. 2003. Marine debris ingestion by southern giant petrels and its potential relationships with fisheries in the southern Atlantic Ocean. *Marine Pollution Bulletin* 46:1513-1515.

Corbin, C.J. and J.G. Singh. 1993. Marine debris contamination of beaches in St. Lucia and Dominica. *Marine Pollution Bulletin* 26:325-328.

Croxall, J.P., S. Rodwell, and I.L. Boyd. 1990. Entanglement in man-made debris of Antarctic fur seals at Bird Island, South Georgia. *Marine Mammal Science* 6:221-233.

Cunningham, D.J. and S.P. Wilson. 2003. Marine debris on beaches of the Greater Sydney Region. *Journal of Coastal Research* 19:421-430.

Day, R.H. and D.G. Shaw. 1987. Patterns in the abundance of pelagic plastic and tar in the north Pacific Ocean, 1976–1985. *Marine Pollution Bulletin* 18:311-316.

Debrot, A.O., A.B. Tiel, and J.E. Bradshaw. 1999. Beach debris in Curacao. *Marine Pollution Bulletin* 38:795-801.

Donohue, M.J. and D.G. Foley. 2007. Remote sensing reveals links among the endangered Hawaiian monk seal, marine debris, and El Niño. *Marine Mammal Science* 23:468-473.

Donohue, M.J., R.C. Boland, C.M. Sramek, and G.A. Antonelis. 2001. Derelict fishing gear in the Northwestern Hawaiian Islands: Diving surveys and debris removal in 1999 confirm threat to coral reef ecosystems. *Marine Pollution Bulletin* 42:1301-1312.

Dufault, S. and H. Whitehead. 1994. Floating marine pollution in the gully on the continental slope, Nova Scotia, Canada. *Marine Pollution Bulletin* 28:489-493.

Edyvane, K.S., A. Dalgetty, P.W. Hone, J.S. Higham, and N.M. Wace. 2004. Long-term marine litter monitoring in the remote Great Australian Bight, South Australia. *Marine Pollution Bulletin* 48:1060-1075.

Erzini, K., C.C. Monteiro, J. Ribeiro, M.N. Santos, M. Gaspar, P. Monteiro, and T.C. Borges. 1997. An experimental study of gill net and trammel net "ghost fishing" off the Algarve (southern Portugal). *Marine Ecology Progress Series* 158:257-265.

Fritt, T.H. 1982. Courtship behavior of the queen snake. *Herpetological Review* 13:72-73.

Frost, A. and M. Cullen. 1997. Marine debris on northern New South Wales beaches (Australia): Sources and the role of beach usage. *Marine Pollution Bulletin* 34:348-352.

Fry, D.M., S.I. Fefer, and L. Sileo. 1987. Ingestion of plastic debris by Laysan albatrosses and wedge-tailed shearwaters in the Hawaiian-Islands. *Marine Pollution Bulletin* 18:339-343.

Furness, R.W. 1985a. Ingestion of plastic particles by seabirds at Gough Island, South Atlantic Ocean. *Environmental Pollution Series A: Ecological and Biological* 38:261-272.

Furness, R.W. 1985b. Plastic particle pollution: Accumulation by procellariiform seabirds at Scottish colonies. *Marine Pollution Bulletin* 16:103-106.

Gabrielides, G.P., A. Golik, L. Loizides, M.G. Marino, F. Bingel, and M.V. Torregrossa. 1991. Man-made garbage pollution on the Mediterranean coastline. *Marine Pollution Bulletin* 23:437-441.

Galgani, F., T. Burgeot, G. Bocquene, F. Vincent, J.P. Leaute, J. Labastie, A. Forest, and R. Guichet. 1995a. Distribution and abundance of debris on the continental shelf of the Bay of Biscay and in Seine Bay. *Marine Pollution Bulletin* 30:58-62.

Galgani, F., S. Jaunet, A. Campillo, X. Guenegen, and E. His. 1995b. Distribution and abundance of debris on the continental shelf of the north-western Mediterranean Sea. *Marine Pollution Bulletin* 30:713-717.

Galgani, F., A. Souplet, and Y. Cadiou. 1996. Accumulation of debris on the deep sea floor off the French Mediterranean coast. *Marine Ecology Progress Series* 142:225-234.

Galgani, F., J.P. Leaute, P. Moguedet, A. Souplet, Y. Verin, A. Carpentier, H. Goraguer, D. Latrouite, B. Andral, Y. Cadiou, J.C. Mahe, J.C. Poulard, and P. Nerisson. 2000. Litter on the sea floor along European coasts. *Marine Pollution Bulletin* 40:516-527.

Galil, B.S., A. Golik, and M. Turkay. 1995. Litter at the bottom of the sea: A sea-bed survey in the Eastern Mediterranean. *Marine Pollution Bulletin* 30:22-24.

Garrity, S.D. and S.C. Levings. 1993. Marine debris along the Caribbean coast of Panama. *Marine Pollution Bulletin* 26:317-324.

Godoy, H., D.M. Furevik, and S. Stiansen. 2003. Unaccounted mortality of red king crab (*Paralithodes camtschaticus*) in deliberately lost pots off northern Norway. *Fisheries Research* 64:171-177.

Golik, A. and Y. Gertner. 1992. Litter on the Israeli coastline. *Marine Environmental Research* 33:1-15.

Gramentz, D. 1988. Involvement of loggerhead turtles with the plastic, metal, and hydrocarbon pollution in the central Mediterranean. *Marine Pollution Bulletin* 19:11-13.

Guillory, V. 1998. Ghost fishing by blue crab traps. *North American Journal of Fisheries Management* 13:459-466.

Harcourt, R., D. Aurioloes, and J. Sanchez. 1994. Entanglement of California sea lions at Los Islotes, Baja California Sur, Mexico. *Marine Mammal Science* 10:122-125.

Haynes, D. 1997. Marine debris on continental islands and sand cays in the far northern section of the Great Barrier Reef Marine Park, Australia. *Marine Pollution Bulletin* 34:276-279.

Hébert, M., G. Miron, M. Moriyasu, R. Vienneau, and P. DeGrace. 2001. Efficiency and ghost fishing of snow crab (*Chionoecetes opilio*) traps in the Gulf of St. Lawrence. *Fisheries Research* 52:143-153.

Henderson, J.R. 2001. A pre- and post-MARPOL Annex V summary of Hawaiian monk seal entanglements and marine debris accumulation in the northwestern Hawaiian Islands, 1982–1998. *Marine Pollution Bulletin* 42:584-589.

Hess, N.A., C.A. Ribic, and I. Vining. 1999. Benthic marine debris, with an emphasis on fishery-related items, surrounding Kodiak Island, Alaska, 1994–1996. *Marine Pollution Bulletin* 38:885-890.

Hofmeyr, G.J.G. and M.N. Bester. 2002. Entanglement of pinnipeds at Marion Island. *South African Journal of Marine Science* 24:383-386.

Hofmeyr, G.J.G., M.N. Bester, S.P. Kirkman, C. Lydersen, and K.M. Kovacs. 2006. Entanglement of Antarctic fur seals at Bouvetoya, Southern Ocean. *Marine Pollution Bulletin* 52:1077-1080.

Imber, M.J. 1996. The food of Cook's petrel *Pterodroma cookii* during its breeding season on Little Barrier Island, New Zealand. *Emu* 96(3):189-194.

Imber, M.J., J.N. Jolly, and M.D. Brooke. 1995. Food of three sympatric gadfly petrels (*Pterodroma* spp.) breeding on the Pitcairn Islands. *Biological Journal of the Linnean Society* 56:233-240.

Kaiser, M.J., B. Bullimore, P. Newman, K. Lock, and S. Gilbert. 1996. Catches in "ghost fishing" set nets. *Marine Ecology Progress Series* 145:11-16.

Katsanevakis, S. and A. Katsarou. 2004. Influences on the distribution of marine debris on the seafloor of shallow coastal areas in Greece (eastern Mediterranean). *Water Air and Soil Pollution* 159:325-337.

Kenyon, K.W. and E. Kridler. 1969. Laysan albatrosses swallow indigestible matter. *Auk* 86:339-343.

Koutsodendris, A., G. Papatheodorou, O. Kougiourouki, and M. Georgiadis. 2008. Benthic marine litter in four Gulfs in Greece, Eastern Mediterranean; abundance, composition and source identification. *Estuarine Coastal and Shelf Science* 77:501-512.

Kusui, T. and M. Noda. 2003. International survey on the distribution of stranded and buried litter on beaches along the Sea of Japan. *Marine Pollution Bulletin* 47:175-179.

Lecke-Mitchell, K.M. and K. Mullin. 1992. Distribution and abundance of large floating plastic in the north-central Gulf of Mexico. *Marine Pollution Bulletin* 24:598-601.

Lee, D.I., H.S. Cho, and S.B. Jeong. 2006. Distribution characteristics of marine litter on the sea bed of the East China Sea and the South Sea of Korea. *Estuarine, Coastal and Shelf Science* 70:187-194.

Madzena, A. and T. Lasiak. 1997. Spatial and temporal variations in beach litter on the Transkei coast of South Africa. *Marine Pollution Bulletin* 34:900-907.

Mallory, M.L., J. Roberston, and A. Moenting. 2006. Marine plastic debris in northern fulmars from Davis Strait, Nunavut, Canada. *Marine Pollution Bulletin* 52:813-815.

Martinez-Ribes, L., G. Basterretxea, M. Palmer, and J. Tintoré. 2007. Origin and abundance of beach debris in the Balearic Islands. *Scientia Marina* 71:305-314.

McCoy, F.W. 1988. Floating megalitter in the eastern Mediterranean. *Marine Pollution Bulletin* 19:25-28.

Merrell, T.R. 1984. A decade of change in nets and plastic litter from fisheries off Alaska. *Marine Pollution Bulletin* 15:378-384.

Moore, C.J., S.L. Moore, M.K. Leecaster, and S.B. Weisberg. 2001. A comparison of plastic and plankton in the North Pacific central gyre. *Marine Pollution Bulletin* 42:1297-1300.

Morris, R.J. 1980a. Plastic debris in the surface waters of the South Atlantic. *Marine Pollution Bulletin* 11:164-166.

Morris, R.J. 1980b. Floating plastic debris in the Mediterranean. *Marine Pollution Bulletin* 11:125.

Moser, M.L. and D.S. Lee. 1992. A fourteen-year survey of plastic ingestion by western North-Atlantic seabirds. *Colonial Waterbirds* 15:83-94.

Nagelkerken, I., G.A.M.T. Wiltjer, A.O. Debrot, and L.P.J.J. Pors. 2001. Baseline study of submerged marine debris at beaches in Curacao, West Indies. *Marine Pollution Bulletin* 42:786-789.

Oigman-Pszczol, S.S. and J.C. Creed. 2007. Quantification and classification of marine litter on beaches along Armação dos Búzios, Rio de Janeiro, Brazil. *Journal of Coastal Research* 23:421-428.

Page, B., J. McKenzie, R. McIntosh, A. Baylis, A. Morrissey, N. Calvert, T. Haase, M. Berris, D. Dowie, P.D. Shaughnessy, and S.D. Goldsworth. 2004. Entanglement of Australian sea lions and New Zealand fur seals in lost fishing gear and other marine debris before and after government and industry attempts to reduce the problem. *Marine Pollution Bulletin* 49:33-42.

Parrish, F.A. and T.K. Kazama. 1992. Evaluation of ghost fishing in the Hawaiian lobster fishery. *Fishery Bulletin* 90:720-725.

Pemberton, D., N.P. Brothers, and R. Kirkwood. 1992. Entanglement of Australian fur seals in man-made debris in Tasmanian waters. *Wildlife Research* 19:151-159.

Pichel, W.G., J.H. Churnside, T.S. Veenstra, D.G. Foley, K.S. Friedman, R.E. Brainard, J.B. Nicoll, Q. Zheng, and P. Clemente-Colón. 2007. Marine debris collects within the North Pacific subtropical convergence zone. *Marine Pollution Bulletin* 54:1207-1211.

Plotkin, P.T., M.K. Wicksten, and A.F. Amos. 1993. Feeding ecology of the loggerhead sea turtle *Caretta caretta* in the northwestern Gulf of Mexico. *Marine Biology* 115:1-5.

Podolsky, R.H. 1989. Entrapment of sea deposited plastic on the shore of a Gulf of Maine Island. *Marine Environmental Research* 27:67-72.

Revill, A.S. and G. Dunlin. 2003. The fishing capacity of gillnets lost on wrecks and on open ground in UK coastal waters. *Fisheries Research* 64:107-113.

Robards, M.D., J.F. Piatt, and K.D. Wohl. 1995. Increasing frequency of plastic particles ingested by seabirds in the subarctic North Pacific. *Marine Pollution Bulletin* 30:151-157.

Ryan, P.G. 1987. The incidence and characteristics of plastic particles ingested by seabirds. *Marine Environmental Research* 23:175-206.

Ryan, P.G. 1988. Intraspecific variation in plastic ingestion by seabirds and the flux of plastic through seabird populations. *Condor* 90:446-452.

Santos, I.R., A.C. Friedrich, and F.P. Barretto. 2005. Overseas garbage pollution on beaches of northeast Brazil. *Marine Pollution Bulletin* 50:782-786.

Santos, M.N., H. Saldanha, M.B. Gaspar, and C.C. Monteiro. 2003. Causes and rates of net loss off the Algarve (southern Portugal). *Fisheries Research* 64:115-118.

Seminoff, J.A., A. Resendiz, and W.J. Nichols. 2002. Diet of East Pacific green turtles (*Chelonia mydas*) in the central Gulf of California, Mexico. *Journal of Herpetology* 36:447-453.

Shaughnessy, P.D. 1980. Entanglement of Cape fur seals with man-made objects. *Marine Pollution Bulletin* 11:332-336.

Shaw, D.G. 1977. Pelagic tar and plastic in the Gulf of Alaska and Bering Sea: 1975. *The Science of the Total Environment* 8:13-20.

Shiomoto, A. and T. Kameda. 2005. Distribution of manufactured floating marine debris in near-shore areas around Japan. *Marine Pollution Bulletin* 50:1430-1432.

Silva-Iñiguez, L. and D.W. Fischer. 2003. Quantification and classification of marine litter on the municipal beach of Ensenada, Baja California, Mexico. *Marine Pollution Bulletin* 46:132-138.

Spear, L.B., D.G. Ainley, and C.A. Ribic. 1995. Incidence of plastic in seabirds from the tropical Pacific, 1984–91: Relation with distribution of species, sex, age, season, year and body weight. *Marine Environmental Research* 40:123-146.

Stefatos, A., M. Charalampakis, G. Papatheodorou, and G. Ferentinos. 1999. Marine debris on the seafloor of the Mediterranean Sea: Examples from two enclosed gulfs in western Greece. *Marine Pollution Bulletin* 38:389-393.

Stewart, B.S. and P.K. Yochem. 1987. Entanglement of pinnipeds in synthetic debris and fishing net and line fragments at San Nicolas and San Miguel Islands, California, 1978–1986. *Marine Pollution Bulletin* 18:336-339.

Storrier, K.L., D.J. McGlashan, S. Bonellie, and K. Velander. 2007. Beach litter deposition at a selection of beaches in the Firth of Forth, Scotland. *Journal of Coastal Research* 23:813-822.

Thiel, M., I. Hinojosa, N. Vásquez, and E. Macaya. 2003. Floating marine debris in coastal waters of the SE-Pacific (Chile). *Marine Pollution Bulletin* 46:224-231.

Tomás, J., R. Guitart, R. Mateo, and J.A. Raga. 2002. Marine debris ingestion in loggerhead sea turtles, *Caretta caretta*, from the Western Mediterranean. *Marine Pollution Bulletin* 44:211-216.

Tschernij, V. and P.O. Larsson. 2003. Ghost fishing by lost cod gill nets in the Baltic Sea. *Fisheries Research* 64:151-162.

Uneputty, P.A. and S.M. Evans. 1997. Accumulation of beach litter on islands of the Pulau Seribu Archipelago, Indonesia. *Marine Pollution Bulletin* 34:652-655.

van Dolah, R.F., V.G. Burrell, and S.B. West. 1980. The distribution of pelagic tars and plastics in the south Atlantic bight. *Marine Pollution Bulletin* 11:352-356.

van Franeker, J.A. 1985a. Plastic ingestion by petrels breeding in Antarctica. *Marine Pollution Bulletin* 19:672-674.

van Franeker, J.A. 1985b. Plastic ingestion by North Atlantic fulmar. *Marine Pollution Bulletin* 16:367-369.

van Franeker, J.A. and P.J. Bell. 1985. Plastic ingestion by petrels breeding in Antarctica. *Marine Pollution Bulletin* 19:672-674.

Velander, K. and M. Mocogni. 1999. Beach litter sampling strategies: Is there a "best" method? *Marine Pollution Bulletin* 38:1134-1140.

Venrick, E.L., T.W. Backman, C. Bartram, C.J. Platt, M.S. Thornhill, and R.E. Yates. 1973. Man-made objects on the surface of the central North Pacific Ocean. *Nature* 241:271.

Vlietstra, L.S. and J.A. Parga. 2002. Long-term changes in the type, but not amount, of ingested plastic particles in short-tailed shearwaters in the southeastern Bering Sea. *Marine Pollution Bulletin* 44:945-955.

Wetzel, L., G. Fillmann, and L.F.H. Niencheski. 2004. Litter contamination processes and management perspectives on the southern Brazilian coast. *International Journal of Environment and Pollution* 21:153-165.

Whiting, S.D. 1998. Types and sources of marine debris in Fog Bay, northern Australia. *Marine Pollution Bulletin* 36:904-910.

Williams, A.T., D.T. Tudor, and P. Randerson. 2003. Beach litter sourcing in the Bristol Channel and Wales, UK. *Water Air and Soil Pollution* 143:387-408.

Willoughby, N.G., H. Sangkoyo, and B.O. Lakaseru. 1997. Beach litter: An increasing and changing problem for Indonesia. *Marine Pollution Bulletin* 34:469-478.

Witherington, B.E. 2002. Ecology of neonate loggerhead turtles inhabiting lines of downwelling near a Gulf Stream front. *Marine Biology* 140:843-853.

Wong, C.S., D.R. Green, and W.J. Cretney. 1974. Quantitative tar and plastic waste distributions in the Pacific Ocean. *Nature* 247:30-32.

Yamashita, R. and A. Tanimura. 2007. Floating plastic in the Kuroshio Current area, western North Pacific Ocean. *Marine Pollution Bulletin* 54:485-488.

Zavala-González, A. and E. Mellink. 1997. Entanglement of California sea lions, *Zalophus californianus californianus*, in fishing in the central-northern part of the Gulf of California, Mexico. *Fisheries Bulletin* 95:180-184.

D

Parties to MARPOL Annex V and Members of Regional Fisheries Management Organizations

Countries and fishing entities that are participants or parties to four key Regional Fisheries Management Organizations (RFMOs): the Inter-American Tropical Tuna Commission (IATTC), the Western and Central Pacific Fisheries Commission (WCPFC), the International Commission for the Conservation of Atlantic Tunas (ICCAT), and the Indian Ocean Tuna Commission (IOTC). These RFMOs are highlighted here because they all have FAD fisheries.

IATTC		WCPFC	
Member countries		**Member countries**	
Colombia	+	Australia	+
Costa Rica		Canada	
Ecuador	+	China	+
El Salvador	+	Chinese Taipei (Taiwan)	
France	+	Cook Islands	
Guatemala	+	European Union*	
Japan	+	Federated States of Micronesia	
Korea, Republic of	+	Fiji	
Mexico	+	France	+
Nicaragua	+	Japan	+
Panama	+	Kiribati	+
Peru	+	Korea, Republic of	+
Spain	+	Republic of the Marshall Islands	+
United States	+	Nauru	
Vanuatu	+	New Zealand	+
Venezuela	+	Niue	
		Palau	
Cooperating nonparties and		Papua New Guinea	
fishing entities		Philippines	+
Belize	+	Samoa	+
Canada		Solomon Islands	+
China	+	Tonga	+
Chinese Taipei (Taiwan)		Tuvalu	+
European Union*		United States	+
		Vanuatu	+
		Cooperating nonparties	
		Belize	+
		Indonesia	
		Participating territories	
		American Samoa	+
		Commonwealth of the	
		Northern Mariana Islands	+
		French Polynesia	+
		Guam	
		New Caledonia	
		Tokelau	
		Wallis and Futuna	

+ Party to MARPOL Annex V, according to the International Maritime Organization (2008).
* While the European Union is not a party to MARPOL Annex V, several EU nations are parties.

ICCAT

Member countries

Angola	+
Albania	+
Algeria	+
Barbados	+
Belize	+
Brazil	+
Canada	
Cape Verde	+
China	+
Côte d'Ivoire	+
Croatia	+
Egypt	+
European Union*	
France (St. Pierre and Miquelon)	+
Gabon	+
Ghana	
Guatemala	+
Guinea, Equatorial	
Guinea, Republic of	+
Honduras	+
Iceland	+
Japan	+
Korea, Republic of	+
Libya	+
Morocco	
Mexico	+
Namibia	+
Nicaragua	+
Nigeria	+
Norway	+
Panama	+
Philippines	+
Russia	+
St Vincent and the Grenadines	+
São Tomé e Principe	+
Senegal	+
South Africa	+
Syria	+
Trinidad and Tobago	+
Tunisia	+
Turkey	+
United States	+
United Kingdom	+
Uruguay	+
Vanuatu	+
Venezuela	+

Cooperating noncontracting parties

Chinese Taipei (Taiwan)	
Guyana	
Netherlands Antilles	+

IOTC

Member countries

Australia	+
Belize	+
China	+
Comoros	+
Eritrea	
European Community	
France	+
Guinea	+
India	+
Indonesia	
Iran	+
Japan	+
Kenya	+
Korea, Republic of	+
Madagascar	+
Malaysia	
Mauritius	+
Oman, Sultanate of	+
Pakistan	+
Philippines	+
Seychelles	
Sri Lanka	+
Sudan	
Tanzania	
Thailand	
United Kingdom	+
Vanuatu	+

Cooperating noncontracting parties

Senegal	+
South Africa	+
Uruguay	+

REFERENCE

International Maritime Organization. 2008. *Status of Conventions by Country*. [Online]. Available: http://www.imo.org/includes/blastDataOnly.asp/data_id%3D22499/status-x.xls [August 4, 2008].

E

Management of Waste and Derelict Fishing Gear[1]

Jenna R. Jambeck, Ph.D.
Department of Civil and Environmental Engineering
University of New Hampshire

OBJECTIVE

Summarize the generally accepted state-of-the-art plastics disposal technologies that may have application to the management of waste fishing gear.

INTRODUCTION

The fishing gear waste stream comprises both nonoperational or otherwise unwanted gear that fishermen wish to dispose of and derelict fishing gear (DFG) that is recovered from the marine environment. To mitigate effects of marine debris pollution, DFG continues to be collected on a worldwide scale. Collection is occurring in remote and isolated areas such as the Northwestern Hawaiian Islands and Dutch Harbor, Alaska, as well as more populated areas like the New England coast. Because of the quantity and composition of this waste stream, management and disposal of the debris can be a challenge. For example, fishing gear is currently made of synthetic materials such as polypropylene, polyethylene, nylon 40, nylon 6, and nylon 66 (Dagli et al., 1990; Timmers et al., 2005). These synthetic materials do not biodegrade (e.g., microbes typically cannot utilize carbon in plastics to create carbon dioxide) and only physically degrade through the changing of the polymers through solar radiation

---

[1] This appendix was prepared at the request of the committee. It has been edited for grammar and style; factual accuracy is the sole responsibility of the author.

and slow thermal oxidation (Gregory and Andrady, 2003), which slowly physically break down the plastics. However, even in smaller pieces, the plastic remains a persistent pollution problem. Though placing the debris in a landfill can be a potentially inexpensive and technologically feasible method of management, because the DFG does not biodegrade, the material occupies landfill airspace indefinitely. In isolated places like Dutch Harbor, this can fill landfills relatively quickly, which creates disposal problems for DFG as well as other wastes.

The purpose of this appendix is to provide potential waste management options for fishing gear. Infrastructure related to waste management at ports and on ships has been previously addressed by the National Research Council (1995), which found that fishing vessels create the third largest quantity of waste (by mass) of the various categories of ships defined (behind recreational and day boats) with a generation rate of 1.85 kg per person per day. Recommendations from the National Research Council included a national infrastructure for the collection and management (recycling and disposal) of old DFG (National Research Council, 1995).

Various options for management of waste on ships are outlined by Hutto (2001); however, this appendix specifically focuses on management of waste fishing gear. After a brief discussion of the composition of fishing gear and waste management, various management and disposal technologies (other than landfilling) for fishing gear, particularly DFG, are described. Table E.1 summarizes these various options and compares them based on their current applicability, feasibility, and requirements.

COMPOSITION OF FISHING GEAR WASTE STREAM

There has been extensive collection and identification of DFG throughout the world (Dagli et al., 1990; Kiessling, 2003; Timmers et al., 2005). Fishing gear was historically composed of natural fibers; nets were composed of cotton flax and hemp (Timmers et al., 2005). When invented, synthetic fibers had many advantages over natural fibers, including durability. This durability also makes DFG persist in the environment. Various studies have been conducted on the composition of DFG—primarily nets. Gregory and Andrady (2003) note that synthetic yarns used in fishing gear include nylon 6, nylon 66, polyester, polyethylene, polypropylene, polyvinyl chloride, polyvinylidene chloride, and polyvinyl acetate. Dagli et al. (1990) collected 1,000 kg of DFG to examine the gear for potential recycling of plastic. Of the 1,000 kg collected, 550 kg represented 49 separate items, including individual nets, individual lines, net combinations, and net/line combinations. The nets were composed of nylon 6, nylon 66, and high-density polyethylene (HDPE); lines were composed of mostly polypropylene. The study also found that nylons could be coated with

TABLE E.1 Summary of Waste Management Options for Fishing Gear

Management Option	Cost (per ton)[a]	Pre-Processing Requirements	Current Technical Feasibility[b]	Feasibility in Remote or Small Location[c]	Scale in US
Landfill	<$30 to >$100	None	3	5*	Full
Recycling	$60	Yes, size reduction (potentially shredding) and densification (bailing)	3	3	Full
Combustion with Energy Recovery	$40 to >$100	Possibly (shredding to make refuse-derived fuel), not needed for mass burn	3	2**	Full
Gasification	Unknown	Yes, 3-inch size	2	3	Demo-commercial
Pyrolysis	Unknown	Yes, <1-inch size, ¾-inch size	2	3	Demo-commercial
Plasma Arc	Unknown	Yes, unknown size	1	3	Demo-research

[a] Costs vary widely and are dynamic. These are current estimates and ranges. Cost is regionally based and market based. Cost also does *not* include transportation.

[b] Scale: 3 = Excellent, proven technology currently in operation and taking waste fishing gear; 2 = Pilot/demonstration technology only in the United States and claims to be able to take waste fishing gear; 1 = Pilot/demonstration technology and ability to accept waste fishing gear unknown.

[c] Scale: 5 = Extremely feasible (in existence); 4 = Very feasible; 3 = Seems feasible, but currently not in existence in remote area; 2 = Not very feasible; 1 = Unfeasible.

* Landfills in remote and small areas are filling up, which is making them less feasible.

** Combustion with energy recovery may not have the required waste input to justify construction of a facility in a small or remote location.

asphaltic and alkyd-type coatings. The nylon and polyethylene did not show signs of degradation (not biodegradation, but polymer change); however, the polypropylene material did show changes resulting from degradation (Dagli et al., 1990). Other references to DFG include a Northwestern Hawaiian Islands database which describes nets as consisting of polypropylene, polyethylene, and nylon 40 (Timmers et al., 2005). Marine debris and DFG in Alaska reportedly include the plastics already referenced, as well as foamed plastics (floats or large blocks) (Bob King, personal communication).

WASTE MANAGEMENT HIERARCHY

The Environmental Protection Agency has developed a waste management hierarchy (Figure E.1). This hierarchy states that waste should be managed in the following order to conserve landfill space resources and potentially reduce carbon emissions: (1) reduce the generation of wastes, (2) reuse and recycle, (3) compost, (4) convert wastes to energy, and (5) landfill. Other portions of the waste management structure (e.g., transportation) should be considered when examining options as well; however, this often necessitates a more detailed assessment such as a life cycle assessment of the waste management options. Other management options that beneficially utilize plastic are logical since it does not biodegrade in a landfill environment and landfill gas cannot be captured from it for beneficial use.

RECYCLING

Recycling of fishing gear has focused on nets (Dagli et al., 1990, 1995; Labib and Maher, 1999). Recycling can take two forms. The synthetic net material can be processed and used as the raw material in the manufacture of the same type of plastic (e.g., HDPE or nylon). In addition, the nets may also be used as fibrous reinforcement for other synthetic materials or used in other constructed compounds, such as asphalt. There are several examples of each of these methods in the literature, as well as a facility that processes and recycles nets currently located in the State of Washington.

Through initial research into the composition and characterization of fishing nets, Dagli et al. (1990) found that extrusion recycling (using the nets as a feedstock) was feasible. Though some of the coatings on the nets needed to be further examined, it was still shown that the nylon 6, nylon 66, and HDPE had not degraded upon exposure to the marine environment and could be recycled into other useful products. The polypropylene line had degraded and was not evaluated for recycling in this research.

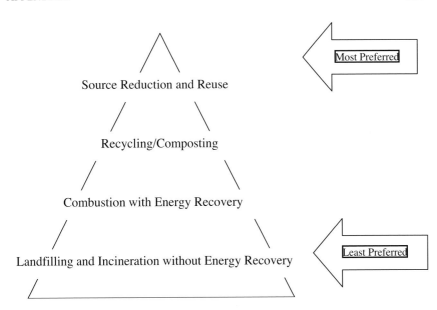

FIGURE E.1 The Environmental Protection Agency's waste management hierarchy (used with permission from the Environmental Protection Agency).

In subsequent work, Dagli et al. (1995) processed nets for recycling and suggested a design for a processing facility for nets. Melt reprocessing was investigated, which encompassed cleaning of the nets, size reduction, melt extrusion and filtering, modification, and injection molding. The nylon 6, nylon 66, and HDPE were blended and utilized in composites that were comparable in mechanical properties to commercially available materials. In companion research, the nets were used as organic fibrous reinforcement in polymeric matrices. Nylon 6 and nylon 66 from the nets were compounded with a thermoplastic polyurethane matrix at temperatures below the melting point of nylons. It was found that good adhesion occurred between the fibers and the thermoplastic polyurethane matrix and also improved physical properties such as stiffness, shore hardness, and abrasion resistance (Dagli et al., 1995).

Fishing net fibers were also tested as an additive to asphalt pavement by the New Jersey Department of Transportation. While carpet and car seat fibers did not work successfully in asphalt pavement, fishing nets did. The fibers from the fishing nets could be uniformly and consistently incorporated into the asphalt mixture without segregation or introduction of excessive air voids (Labib and Maher, 1999). It is not known if further research or implementation of this technology continued after this initial research study.

The technical feasibility in the literature has been put into practice by Skagit Steel of Burlington, Washington. Skagit Steel processes fishing nets into products meeting the specifications of various end users. Their processing includes cleaning, densification (bailing), and shipping; none of the plastics are melted or extruded onsite. Skagit Steel reported processing approximately 500 tons of netting in 2007 (Lois Young, personal communication); this included 10 tons of mixed unsorted marine debris, primarily DFG, collected in Dutch Harbor, Alaska, with a tipping fee (cost to customer) of approximately $60 per ton (Bob King, personal communication). All forms of nets are accepted and the tipping fee is based on the condition of the nets (e.g., amount of organic contamination). The majority of the nets accepted are clean, but nets with some contaminants such as organics (e.g., algae, mussels, and other fouling organisms) and metals (e.g., lead lines) are also accepted. When contaminants exist, this can require manual sorting of the netting material (requiring the higher tipping fee), while clean nets can often be processed mechanically. Nets must be processed to meet various specifications determined by the end use. Most end users require that the material they receive has no organic life or metals. The end products of this process are reportedly recycled into new plastics (feedstock), upholstery, heat-resistant bearings, and plastic lining material (Lois Young, personal communication).

The cost of operating a recycling facility for plastic fishing gear is similar to the costs of operating other material recovery facilities. Construction and demolition debris is likely the most similar material that is currently processed. The critical aspect of recycling is that enough throughput exists so that sufficient product can be sold for income. The available income directly relates to the market for the products, which is often variable and can make operations difficult. However, diversifying the materials processed by a single facility helps to offer a variety of products from processing that could provide income. Because of market fluctuations, it would make sense for an existing recycling facility to incorporate processing of fishing nets into their operations—or for a new facility to process more than just fishing nets (e.g., also accept metals). For locations that have a consistent stream, a fishing-gear-only recycling facility might also be feasible, though a specific cost analysis would be needed. In addition, processing facilities located remotely would have to take into account the transport of the densified (bailed) nets, unless there are local end users for the plastic. After processing, the material would likely be classified as a product and not a waste, allowing it to be shipped with other goods being shipped between states or overseas; for example, China currently has a market for plastic recycling.

THERMAL TREATMENT TECHNOLOGIES

Thermal technologies involve temperature changes to convert waste materials into useable products (e.g., organics into energy). They include the breakdown of the input materials into their elemental forms at a particular efficiency based on operational characteristics. Also, some thermal treatment facilities include various combinations of the technologies outlined in this section. There can be some confusion over the difference between gasification and pyrolysis technologies. Sometimes the production of gas in the pyrolysis process is referred to as gasification and sometimes the transformation of the organic materials into their elemental form is referred to as pyrolysis. However, for the purposes of this paper, the technologies are defined separately as described in subsequent sections.

Combustion with Energy Recovery

Facilities that combust waste and recover the energy are often called waste-to-energy (WTE) facilities. Of the 251 million tons of waste generated in the United States in 2006, 31.4 tons (12.5 percent) were combusted with energy recovery (Environmental Protection Agency, 2007). WTE technologies are proven and facilities are operated by both public entities and private companies throughout the United States. WTE reduces the volume of solid waste; for plastics, volume is typically reduced by 90 percent (Environmental Protection Agency, 2007).

WTE technologies have been used for managing the fishing gear waste stream. However, the location of these facilities is important. Remote locations such as Dutch Harbor, Alaska, may not have close access to a facility and transportation of waste is expensive. In the Northwestern Hawaiian Islands, the fishing gear collected as a part of the Ghost Net Identification study was combusted with energy recovery at Honolulu Power (HPower), operated by Covanta. Currently, the Nets to Energy project continues to operate and deliver nets to HPower. HPower operates a refuse-derived fuel (RDF) combustion facility with energy recovery. Through size reduction and some sorting, an RDF facility preprocesses waste into a fuel that has optimum energy content and efficient combustion. Since HPower uses RDF, the fishing gear and nets must also be preprocessed before combustion. The nets were historically processed at Hawaii Metal Recycling Company and are now currently processed at Schnitzer Steel Hawaii, which size-reduces the nets for use at HPower. In 2003, 111 metric tons of fishing gear were utilized at HPower, creating energy that equated to powering 42 homes in Oahu for one year (Timmers et al., 2005; Yates, 2007). In the first year of the Nets to Energy program, 11 metric tons of debris were managed (Timmers et al., 2005).

In Massachusetts and New Hampshire, fishing gear collection and management programs have commenced with both Covanta and Wheelabrator (Covanta Energy, 2008; New Hampshire Sea Grant, 2008). The facilities for these two projects are mass burn facilities (i.e., no preprocessing of the waste is required), so size reduction of the nets is not needed before combustion with energy recovery. Fouling organisms and other organics should not be a problem at a WTE facility because they will also combust to create energy. Metal is undesirable at a WTE facility, but small amounts can be tolerated. So far, the nets collected in New England have not needed source separation for metals; however, based on feedback from facility operators, this could be required depending on the amounts of metals and the facility design and operation.

For a WTE facility to remain cost effective, it needs a steady stream of waste to operate and create electricity. Because of the investment in air pollution control systems, which can be large and expensive, facilities taking small amounts of waste are not likely to be economically feasible, nor are those located in remote areas without consistent waste material inputs. A site-specific waste flow analysis and design would be needed to determine economic feasibility. However, a WTE facility could take more waste than just fishing gear; it could take all (or a portion of) municipal solid waste generated in a specific area as well. Permitting a WTE facility, which is conducted by each individual state, is an extensive process as it often requires solid waste, air, water, and stormwater permits, as well as potential permits for land use. WTE facilities have also historically been controversial and public comment and input is required before permitting, construction, and operation.

Gasification

Gasification is the conversion of organic waste into its basic building blocks (carbon monoxide and hydrogen) with a small amount of oxygen input in the process. Gasification includes the partial oxidation of organics into a high-temperature gas in a reducing atmosphere and often uses air steam or oxygen as the gasification agent (Ray and Thorpe, 2007). The exothermic reaction between the carbon and the oxygen can provide the heat energy required to drive the process. The beneficial output is a flammable synthesis gas, or syngas, primarily composed of hydrogen, carbon monoxide, carbon dioxide, methane, and also nitrogen if air is used as the gasification agent (Ray and Thorpe, 2007). Gasification is an exothermic process and, since a small amount of oxygen (or air) is used, some carbon from the waste is lost as carbon dioxide instead of being converted into fuel, making gasification less efficient at conversion than pyrolysis (Ray and Thorpe, 2007). While there are some potential advantages over incineration (e.g., potentially less dioxin and furan formation because of the reducing environment and temperature), the syngas can still contain impurities and requires extensive gas cleaning before being used beneficially (Ray and Thorpe, 2007). While literature is not available specifically for gasification of fishing gear, gasification as a process to convert plastics is proven (Pinto et al., 2002; Megan Feldt, personal communication) and several viable commercial processes are available; because of this, gasification was chosen as the preferred technology by the Sustainable Plastics to Olefins Recycling Technology project in the United Kingdom in 2007 (Ray and Thorpe, 2007).

Ze-gen is a commercially operating demonstration gasification facility in Massachusetts. Ze-gen gasifies up to 10 tons per day of construction and demolition residual material (which also includes some mixed plastics). The gasification process utilizes molten bath technology to produce syngas (primarily carbon monoxide and hydrogen). This syngas will eventually be used as fuel to generate electricity in a full-scale facility (Megan Feldt, personal communication). Slag is produced as a byproduct in the gasification process and is proposed to be used as construction aggregate. The Ze-gen facility has been site assigned by the City of New Bedford and permitted by the Massachusetts Department of Environmental Protection to handle, process, and transfer up to 1,500 tons per day of construction and demolition material, municipal solid waste, and scrap tires. The demonstration test facility began operating in October of 2007. Ze-gen's full-scale facility could accept fishing gear waste (primarily plastics) if it were size reduced to 3-inch by 3-inch pieces (Megan Feldt, personal communication).

Gasification can accept a relatively diverse input stream (less diverse than WTE, but more diverse than a plastics-to-fuel conversion) and this diversity could help a facility to be sited where a more specific process (plastics-to-fuel) would not have a consistent waste input stream. Gasification can take scrap tires and wood (biomass). In addition, the scale can be smaller than that of a WTE facility because the investment in air pollution equipment is not of the same scale. However, gas purification is needed and the infrastructure for this must be available for beneficial use of the gas. Purification must either be at the facility itself or at a location within transport distance. A site-specific waste, cost, and energy analysis would be required to determine economic feasibility.

Pyrolysis

Pyrolysis is the thermal conversion of materials (e.g., waste) in the absence of oxygen. Pyrolysis is an endothermic process requiring energy *input*, but it is very efficient at conversion (more so than gasification) (Ray and Thorpe, 2007). Products of pyrolysis can include gases (e.g., carbon monoxide, hydrogen syngas), liquids and waxes (e.g., fuels, oil, diesel), and solid residue (e.g., char, coke, and carbon black) (Ludlow-Palafox and Chase, 2001). Research indicates that the products depend significantly on the pyrolysis process employed and variations in reactor vessel, retention time, and temperature, as well as other process details (Ludlow-Palafox and Chase, 2001; Kim et al., 2005a, b).

In Korea, where DFG (composed of nylon 6, polyethylene, and polypropylene) has historically been placed in landfills, pyrolysis has been investigated as an alternative waste management strategy (Kim et al., 2005a, b). Kinetic tests using thermogravimetric analysis on nylon 6 show that gas, oil, and a small amount of coke are produced upon pyrolysis. The yield of gas compounds increased with the increase of reaction time (Kim et al., 2005a). It is also reported that higher process temperatures lead to higher yields of gases (Ludlow-Palafox and Chase, 2001). In Korea, the pyrolyzed oil from nylon 6 contained both nitrogen and oxygen (Kim et al., 2005b). Without any further processing, this oil would produce nitrogen oxides upon combustion; however, technologies exist to control postcombustion nitrogen oxide emissions.

Pyrolysis requires an energy input and must produce net energy for economically feasible and sustainable operation. With the price of oil continuing to rise, the conversion of plastics to fuel is quickly evolving into an applicable technology. Demonstration and commercially operated facilities have existed overseas. A 2.5-ton-per-day waste polystyrene processing plant exists in Okayama, Japan. The plastic is treated with pyrolysis to produce liquid oil similar to kerosene (Klean Industries, 2006). Currently, in the United States, Plas2Fuel of Kelso, Washington, is employing a third-generation commercial-scale process which has been operating since the beginning of 2008 (Kevin DeWhitt, personal communication). The first application of the technology has not been tried yet, but the target is to initially convert agricultural waste plastic (mixed). Plas2Fuel reports that it is not intending to compete with segregated plastics recycling because separated and segregated plastics have a higher value. As a company, it is targeting mixed plastics only. In terms of technical feasibility, plastics can be mixed with organic and inorganic materials; however, anything not turned into fuel, including metal contaminants, leads to a greater quantity of byproducts (Kevin DeWhitt, personal communication).

The plastic feedstock for the Plas2Fuel process does not have to be sorted or washed, but it must be size reduced to smaller than 1 inch. The Plas2Fuel process produces synthetic oil, which would need further refinement to make a viable fuel or lubricant. As an example, outputs from the Plas2Fuel process might be 7 percent carbon (black solid), 3 percent hydrochloric acid, and 90 percent oil and light gases. Both the carbon and acid wastewater are byproducts. The carbon black byproduct could have a beneficial use and Plas2Fuel is looking for secondary uses. The wastewater might have a secondary use as well, but, because of the acidic strength, a secondary user might be difficult to find. The level of metals in the solids generated depends on the feedstock. Both cadmium and lead are used as modifiers in plastics and can impact the composition of the solid residue. For example, if enough metal were present, the resulting solid might not pass the Environmental Protection Agency's Toxicity Characteristic Leaching Procedures, qualifying it as a hazardous waste; however, this is reportedly rare (Kevin DeWhitt, personal communication).

The process is new for the State of Washington and is not exempt from permits, but it is not currently regulated under any permits except for air quality. The technology itself is not specifically prohibited. Since the process operates under a vacuum, there are no air emissions. However, because of the light gas recycling, the air regulation issues can be complex; however, Plas2Fuel is meeting all applicable regulations at this time (Kevin DeWhitt, personal communication).

Another company, TSphere Energy of Hawaii, is marketing a process developed by Adia Japan Co., Ltd. The process is known as plastic fuel conversion (Kate Butterfield, personal communication). Based on product literature provided by TSphere, the process has a 95 percent oil recovery and then utilizes 7 percent to operate a generator to power the process. The plastic waste must be size reduced to three-quarters of an inch. The fuel produced is a #1 heavy oil. The plastic is first liquefied (with heat) and then is thermally decomposed without the use of a catalyst. The gasified material is cooled (condensed) and stored for use. Varying plastic fuel converters offered by TSphere Energy are reportedly processing 0.9–3 tons per day (290–1,200 tons per year) with 4–6-month manufacturing and setup lead times (Kate Butterfield, personal communication).

Since there are no full-scale operating facilities in the United States, as with any conversion process outlined in this paper, in order to justify the investment in a pyrolysis conversion facility, a constant feedstock (or a large stockpile) of material would need to be available. It is also not clear what the market is for the varying products produced from the process (i.e., oil, char, and wastewater). An individual assessment and cost-benefit analysis would be needed before considering a facility.

Plasma Arc Furnace and Vitrification

Plasma arc heaters are electric arc heaters (need electrical energy) that include the presence of an ionized gas (plasma) such as hydrogen (reducing), oxygen (oxidizing), or argon (inert) (Electric Power Research Institute, 1991; National Research Council, 1996). Plasma arc heaters operate at extremely high temperatures—core temperatures of 7,200–36,000°F and gas temperatures of 3,600–5,400°F—and have been used in many applications, including heating and melting of metals, reclaiming of metals, smelting of ores, and treatment of dusts and various wastes (Electric Power Research Institute, 1991; Chua et al., 2006). While commercial technology is available for treatment of wastes with energy input, plasma arc is being investigated for waste treatment with energy recovery as well. The energy recovery is similar to that of pyrolysis and gasification—recovery of the basic building blocks of the waste stream itself (organics through syngas). The inorganic portion of the waste is vitrified into a glass-like slag material that could potentially be beneficially utilized (e.g., in construction). Various entities, including the U.S. Department of Defense and cities and counties in Florida (e.g., Tallahassee, St. Lucie County), are investigating plasma arc use (National Research Council, 1996; Shifler and Wong, 1997). Both a ship-based system (Plasma Arc Waste Destruction System [PAWDS]) and a mobile system (Plasma energy Pyrolysis system [PEPS®]) have been developed for the U.S. Department of Defense. Plasma vitrification has been explored for the management of noncombustible fiber-reinforced plastic, gill nets, and waste glass in Taiwan (Chua et al., 2006).

The ship-based plasma arc system, PAWDS, was developed cooperatively with PyroGenesis and the U.S. Navy. Work was ongoing with the U.S. Navy as of a 1997 report on Material Considerations for the Navy Shipboard Waste Destruction System. For reasons not stated in that report, plastics were not considered a material to be treated by PAWDS at that time. According to PyroGenesis, PAWDS has been successfully installed on a Carnival Cruise Lines Ship (PyroGenesis, 2008). The system can be designed for 0.1–15 tons per day capacity and energy recovery is optional. Size reduction of the waste is required and the system includes a waste shredder. PAWDS produces a sand-like ash which can be off-loaded in port or disposed of at sea (PyroGenesis, 2008).

PEPS® was developed and demonstrated in cooperation with the U.S. Army by Enersol Technologies, Inc. The first phase of research was a stationary PEPS® to evaluate reliability, maintainability, and overall effectiveness in destroying problematic waste streams on a commercial scale. The operational testing of a 10-ton-per-day facility was completed in 1999 (EnerSol Technologies, Inc., 2008). System and environmental performance was evaluated by an independent testing laboratory that was responsible for sampling and analysis of process emissions and byproducts (EnerSol Technologies, Inc., 2008). Since the development of the stationary PEPS®, a mobile PEPS® has been under development.

While there are no current operating plasma arc facilities in the United States, there is a facility in Japan that accepts approximately 165 tons per day of automobile shredder residue as fuel, producing approximately 8 megawatts of electric power. The facility could accept up to 330 tons per day of municipal solid waste (Vaidyanathan et al., 2007). Just like the evaluations for plasma arc facilities taking place currently (e.g., Florida), a site-specific evaluation of the waste, cost, and energy production would be required before determining the feasibility of a plasma arc facility for fishing gear or other wastes.

REFERENCES

Chua, J.P., Y.T. Chena, T. Mahalingam, C.C. Tzeng, and T.W. Cheng. 2006. Plasma vitrification and re-use of non-combustible fiber reinforced plastic, gill net and waste glass. *Journal of Hazardous Materials*. B138:628-632.

Covanta Energy. 2008. *Press Release: Covanta Energy and NOAA Join Together with the National Fish and Wildlife Foundation to Launch Fishing for Energy Program*. [Online]. Available: http://www.reuters.com/article/pressRelease/idUS137343+05-Feb-2008+BW20080205 [May 19, 2008].

Dagli, S.S., A. Patel, and M. Xanthos. 1990. Reclaiming and recycling of discarded plastic fishing gear. *Polymeric Materials Science and Engineering: Proceedings of the ACS Division of Polymeric Materials Science and Engineering* 63:1024-1028.

Dagli, S.S., S. Dey, R. Tupil, and M. Xanthos. 1995. Value-added blends and composites from recycled plastic fishing gear. *Journal of Vinyl and Additive Technology* 1:195-200.

Electric Power Research Institute. 1991. *Techcommentary: Plasma Arc Technology*. Center for Materials Production, Carnegie Mellon Research Institute.

EnerSol Technologies, Inc. 2008. *EnerSol Technologies, Inc.: PEPS® and PEGS™ Plasma Enhanced Systems*. [Online]. Available: http://www.enersoltech.com [May 22, 2008].

Environmental Protection Agency. 2007. *Municipal Solid Waste in the United States: 2006 Facts and Figures*. Solid Waste and Emergency Response, Washington, DC.

Gregory, M.R. and A.L. Andrady. 2003. Plastics in the marine environment. In *Plastics and the Environment*, Andrady, A.L. (ed.). John Wiley & Sons, Inc., New York.

Hutto, L.B. 2001. *A Comprehensive Guide to Shipboard Waste Management*. MTS/IEEE Conference and Exhibition, Honolulu, HI.

Kiessling, I. 2003. *Finding Solutions: Derelict Fishing Gear and Other Marine Debris in Northern Australia, National Oceans Office and Department of the Environment and Heritage*. Charles Darwin University, National Oceans Office, Canberra, Australia.

Kim, S., J. Jeon, Y. Park, and K. Kim. 2005a. Thermal pyrolysis of fresh and waste fishing nets. *Waste Management* 25:811-817.

Kim, S., B. Chun, and J. Jeon. 2005b. Pyrolysis kinetics and characteristics of the mixtures of waste ship lubricating oil and waste fishing rope. *Korean Journal of Chemical Engineering* 22(4):573-578.

Klean Industries. 2006. *Focused on Renewable Energy, Resource Recovery and Recycling*. [Online]. Available: http://www.kleanindustries.com/s/Home.asp [May 22, 2008].

Labib, M. and A. Maher. 1999. *Recycled Plastic Fibers for Asphalt Mixtures*. [Online]. Available: http://www.cait.rutgers.edu/finalreports/FHWA-NJ-2000-004.pdf [July 16, 2008].

Ludlow-Palafox, C. and H.A. Chase. 2001. Microwave-induced pyrolysis of plastic wastes. *Industrial and Engineering Chemistry Research* 40(22):4749-4756.

National Research Council. 1995. *Clean Ships, Clean Ports, Clean Oceans: Controlling Garbage and Plastic Wastes at Sea*. National Academy Press, Washington, DC.

National Research Council. 1996. *Shipboard Pollution Control: U.S. Navy Compliance with MARPOL Annex V*. National Academy Press, Washington, DC.

New Hampshire Sea Grant. 2008. *Marine Debris to Energy.* [Online]. Available: http://cecf1.unh.edu/debris [May 27, 2008].

Pinto, F., C. Franco, R.N. Andre, M. Miranda, I. Gulyurtlu, and I. Cabrita. 2002. Co-gasification study of biomass mixed with plastic wastes. *Fuel* 81:291-297.

PyroGenesis. 2008. *Advanced Waste-to-Energy Plasma Systems.* [Online]. Available: http://www.pyrogenesis.com/index.asp [July 1, 2008].

Ray, R. and R.B. Thorpe. 2007. A comparison of gasification with pyrolysis for the recycling of plastic containing wastes. *International Journal of Chemical Reactor Engineering* 5(1):A85.

Shifler, D.A. and C.R. Wong. 1997. *Material Considerations for the Navy Shipboard Waste Destruction System, Survivability, Structures and Materials Directorate Technical Report.* Naval Surface Warfare Center, West Bethesda, Maryland.

Timmers, M.A., C.A. Kistner, and M.J. Donohue. 2005. *Marine Debris of the Northwestern Hawaiian Islands: Ghost Net Identification.* Hawaii Sea Grant Publication.

Vaidyanathan, A., J. Mulholland, J. Ryu, M.S. Smith, and L.J. Circeo, Jr. 2007. Characterization of fuel gas products from the treatment of solid waste streams with a plasma arc torch. *Journal of Environmental Management* 82:77-82.

Yates, L. 2007. *Nets to Energy: The Honolulu Derelict Net Recycling Program.* [Online]. Available: http://www.csc.noaa.gov/cz/2007/Coastal_Zone_07_Proceedings/PDFs/Thursday_Abstracts/3315.Yates.pdf [August 1, 2008].